CEREBRUM 2015

Cerebrum 2015

Emerging Ideas in Brain Science

Bill Glovin, Editor

New York

Published by Dana Press, a Division of the Charles A. Dana Foundation, Incorporated

Address correspondence to:
Dana Press
505 Fifth Avenue, Sixth Floor
New York, NY 10017

New York, NY 10017
DANA is a federally registered trademark.
Printed in the United States of America
ISBN-13: 978-1-932594-59-1
ISSN: 1524-6205
Book design by Bruce Hanson at EGADS (egadsontheweb.com)
Cover illustration by William Hogan

CONTENTS

KEY TO COVER ILLUSTRATION

HORIZONTAL (FROM LEFT):

Schizophrenia (Page 73) is the focus of *Cerebrum*'s July article, "Schizophrenia: Hope on the Horizon."

DNA (Page 37) is the acronym for deoxyribonucleic acid, referenced in *Cerebrum*'s April article, "The Darkness Within: Individual Differences in Stress." The author points to seminal work that showed that the activation of dopamine receptors stimulates a cascade of events that ultimately lead to changes in the rate of DNA transcription initiation and alterations in gene expression.

Neuron—a brain cell responsible for the transmission of nerve impulses—is referenced in almost every *Cerebrum* article and book review in 2015.

PIL (Page 37) is the acronym for Purpose in Life, the focus of *Cerebrum*'s June article, "New Movement in Neuroscience: A Purpose-Driven Life."

Genes is referenced is several *Cerebrum* articles in 2015. A better understanding of genes, which are the basic unit of inheritance, holds enormous potential for the treatment of many brain-related disorders.

MS is the acronym for muscular dystrophy, which is a major area of research of Adam Kaplin, co-author of *Cerebrum*'s June article, "New Movement in Neuroscience: A Purpose-Driven Life." Kaplin, M.D., Ph.D., is the chief psychiatric consultant to the Johns Hopkins Multiple Sclerosis and Transverse Myelitis Centers and is on the board of medical advisors to the Montel Williams MS Foundation and the Nancy Davis MS Foundation.

FDA (Pages 91 and 133) is the acronym for the Food and Drug Administration (FDA), whose role in drug research is addressed in "No End in Sight: The Abuse of Prescription Narcotics," *Cerebrum*'s September article, and in *Power Foods for the Brain*, a *Cerebrum* book reviewed by David O. Kennedy, Ph.D.

Tau's (Page 51) role as a protein is the focus of "Tau-er of Power," *Cerebrum*'s May article.

Cannabis (Page 3) and its tie to schizophrenia in the developing brain is the focus of "Appraising the Risks of Reefer Madness," *Cerebrum*'s January article.

MRI is the acronym for magnetic resonance imaging, a test that uses a magnetic field and pulses of radio wave energy to make pictures of organs and structures inside the body. A form of MRI technology that measures brain activity is functional neuroimaging, or (f)MRI—used to diagnose various cognitive disorders and for further research in several of the year's *Cerebrum* articles.

Snort (Page 91), which is the way heroin and cocaine and various painkillers are ingested, is addressed in "No End in Sight: The Abuse of Prescription Narcotics," *Cerebrum*'s September article.

NIH is the acronym for National Institutes of Health, the major funding body for much of the research cited in almost every *Cerebrum* article in 2015. Under its umbrella is the National Institute of Mental Health, the largest scientific organization in the world dedicated to research focused on the understanding and treatment of mental illnesses.

MDD (Page 81) is the acronym for Manic Depressive Disorder, the focus of "The Holy Grail of Psychiatry," *Cerebrum*'s August article.

Storytelling (Page 15), which can cause oxytocin release and affect attitudes, is the focus of "Why Inspiring Stories Make Us React: The Neuroscience of Narrative," *Cerebrum*'s February article.

VERTICAL (FROM LEFT)

DSM (Page 103), which is also known as the *Diagnostic and Statistical Manual of Mental Disorders* (DSM-5), is the standard classification of mental disorders used by mental health professionals in the United States. In the 2013 edition DSM-5, binge-eating disorder, the focus of *Cerebrum*'s October article, was added.

Stress (Page 37) is the focus of "The Darkness Within: Individual Differences in Stress," *Cerebrum*'s April article. It is also referenced in *The Teenage Brain*, a *Cerebrum* book reviewed by Marisa M. Silveri, Ph.D.

NB (Page 133) references the initials of Neal Barnard, the author of *Power Foods for the Brain*, a *Cerebrum* book reviewed by David O. Kennedy, Ph.D.

CTE (Page 51) is the acronym for chronic traumatic encephalopathy, which is addressed in "Tau-er of Power," *Cerebrum*'s May article.

FJ (Page 151) references the initials of Frances Jensen, the author of *The Teenage Brain*, a *Cerebrum* book reviewed by Marisa M. Silveri, Ph.D.

RNA (Page 51) is the acronym for ribonucleic acid, a polymeric molecule implicated in various biological roles in coding, decoding, regulation, and expression of genes. RNA is referenced in "Tau-er of Power," *Cerebrum*'s May article.

Oxy (Page 91) is slang for Oxycontin, a synthetic opiate that is discussed in "No End in Sight: The Abuse of Prescription Narcotics," *Cerebrum*'s September article.

Mountcastle (Page 27) is the surname of pioneering neuroscientist Vernon B. Mountcastle, M.D., who died on January 9, 2015. He is remembered by two colleagues, Mahon R. DeLong, M.D., and Guy McKhann, M.D., in "Vernon Remembered," *Cerebrum*'s March article.

LC (Page 139) references the initials of Pulitzer Prize winner Leon Cooper, the author of *Science and Human Experience: Values, Culture, and the Mind*, a *Cerebrum* book reviewed by Gary S. Lynch, Ph.D.

PSG (Page 73) is the acronym for the Psychiatric Genomics Consortium, a worldwide group that shares research information on nine mental health disorders, including schizophrenia. The group was part of the focus of *Cerebrum*'s July article, "Schizophrenia: Hope on the Horizon."

Replication (Page 113) is the focus of "Failure to Replicate: Sound the Alarm," *Cerebrum*'s November article about the most comprehensive investigation ever done about the rate and predictors of reproducibility in social and cognitive sciences.

Aging (Page 121) and its impact on cognitive function is the focus of "Cognitive Skills and the Aging Brain: What to Expect," *Cerebrum*'s December article.

AD (Pages 51, 61, and 121) is the acronym for Alzheimer's disease. Several *Cerebrum* articles in 2015 addressed AD, including "Tau-er of Power," *Cerebrum*'s May article; "New Movement in Neuroscience: A Purpose-Driven Life," *Cerebrum*'s June article; and "Cognitive Skills and the Aging Brain: What to Expect," *Cerebrum*'s December article.

MG (Page 145) references the initials of Michael Gazzaniga, the author of *Tales from Both Sides of the Brain: A Life in Neuroscience*, a *Cerebrum* book reviewed by Theodor Landis, Ph.D.

BED (Page 103) is the acronym for binge-eating disorder, the focus of "The Binge and the Brain," *Cerebrum*'s October article.

DEA (Page 91) is the acronym for the Drug Enforcement Administration, whose role is referenced in "No End in Sight: The Abuse of Prescription Narcotics," *Cerebrum*'s September article.

Foreword

By Alan I. Leshner, Ph.D.

Alan I. Leshner, Ph.D., is chief executive officer, emeritus, of the American Association for the Advancement of Science (AAAS) and former executive publisher of the journal *Science*. He served at AAAS/*Science* from 2001 to 2015. Before this position, Leshner was director of the National Institute on Drug Abuse at the National Institutes of Health (NIH). He also served as deputy director and acting director of NIH's National Institute of Mental Health and in several roles at the National Science Foundation. Before joining the government, Leshner was professor of psychology at Bucknell University. Leshner is an elected fellow of AAAS, the American Academy of Arts and Sciences, the National Academy of Public Administration, and many other professional societies. He is a member and served on the governing council of the National Academy of Medicine (previously the Institute of Medicine) of the National Academies of Sciences, Engineering and Medicine. He was appointed by President George W. Bush to the National Science Board in 2004, and then reappointed by President Barack Obama in 2011. Leshner received M.S. and Ph.D. degrees in physiological psychology from Rutgers University and an A.B. in psychology from Franklin & Marshall College. He has been awarded seven honorary doctor of science degrees.

SCIENTIFIC KNOWLEDGE IS ADVANCING at an accelerating pace, with great attendant benefits for humankind. This is true in virtually all fields but seems particularly true for neuroscience, where we have been noting, every few years throughout the 40 years that I have been in the field, that we have learned more about the brain in the last decade than in all of recorded history. And it is continuously true. The advances described in this year's *Cerebrum* anthology are examples of that kind of great progress, and also illustrate some fundamental and general principles about the nature of science as an enterprise.

Progress in science, including in neuroscience, typically occurs in two ways. Most progress is what is termed "incremental." Here findings systematically build on each other, adhering more or less to what most often is thought of as the scientific method: hypotheses are deduced from previously discovered facts; those hypotheses are then systematically tested, usually through formal experimentation, and then the results of those studies are integrated into either existing or new theories. Practicing scientists know, of course, that they rarely are this systematic in their work, but the fundamental notion of "incremental" progress describes well one path to scientific advance, whatever the actual sequence of thought and practice.

The second way in which progress occurs is much more unusual, and is often thought of as "disruptive" or "transformative." In this case a new finding or phenomenon is totally unanticipated or wholly outside the incremental progression and can radically change our understanding or the way we view things. These kinds of advances have at times been called "scientific revolutions," although not all transformative findings cause that level of disruption. It is worth mentioning that these kinds of transformative findings or perspectives are very often met with skepticism or even outright ridicule, since they challenge the conventional wisdom. Fortunately, the scientists responsible for such advances have persisted, to the benefit of the entire field. The progress in neuroscience summarized in this year's *Cerebrum* anthology illustrates both routes of scientific advance.

One important additional characteristic of the scientific enterprise illustrated in this anthology is that it is a self-correcting system. Progress occurs in all fields in "fits and starts," with most errors and outright mistakes eventually being caught or corrected as scientists review, replicate, and evaluate each other's work. This is a central element in how we refine our theories and fact-base. Today's great idea or discovery may be disproved tomorrow, or an idea deemed way off the mark today may turn out to be a great one tomorrow. For scientists, our self-correcting system is normal business. As a side note, it is unfortunate that some policymakers, members of the media, and others in the general public can misunderstand the way scientific progress comes about and overreact when corrections occur. The scientific community can help with this problem by working to educate the public about the nature of our enterprise and how progress comes about, as well as about specific findings. Chapters in this anthology illustrate these trends and principles well.

The book review by **Marisa Silveri** of **Frances Jensen** and **Amy Ellis Nutt**'s book on the teenage brain and the article by **Diane Howieson** on cognitive skills and the aging brain show how advances in science can revise common wisdom and long-held views about the normal course of lifespan development and the brain mechanisms underlying cognitive and emotional behavior. It has only been within the last 15-20 years that we have begun to understand the dramatic brain changes that occur over the course of adolescence and how they might help account for characteristic adolescent behavior patterns. Before then, many people believed that the brain was fully developed by the onset of adolescence and that there were relatively few changes thereafter. Howieson's paper gave me personal hope by pointing out that not all changes in the aging brain have negative consequences. She argues that the proverbial "wisdom of the aged" may have some basis in age-related changes in brain structure and function, and provides evidence to suggest the underlying mechanisms.

Kenneth Kosik's piece on the role of tau and "tauopathies" in dementias teaches much about dementias but also includes an excellent discussion of prions and their potential role in disease processes. Prions are a great example of a discovery that took very long to take hold. When Stanley

Prusiner first introduced the concept of prions as a central mechanism in disorders such as Creutzfeldt-Jakob disease, it was met with great resistance. Prusiner later was awarded the Nobel Prize for his seminal, transformative work. I am sure we all can think of other examples of advances that were first met with resistance.

Patrick Sullivan's article on schizophrenia and psychiatric genomics illustrates how progress occurs by fits and starts, and tells a very interesting story of how modern psychiatric geneticists are working to tackle a problem that seems obvious on the surface but has been nearly impenetrable so far: understanding the hereditary component of mental illnesses. The search for genes affecting mental disorders has been ongoing for almost as long as we have known that mental illnesses tend to run in families. Earlier approaches yielded false starts—at times with great public fanfare and overblown media attention. Now, the Psychiatric Genomics Consortium is using the most modern genetic technologies to study hundreds of thousands of subjects, looking at relationships among arrays of genes, diagnoses and symptom clusters. Understanding which genes are involved in which illnesses or symptoms is critical, of course, not only for understanding the pathophysiology of mental illnesses but also for identifying biological targets for new medication. I am optimistic that this approach will bring the full power of genetics and genomics to bear on this problem on a scale that may, at last, yield some real understanding.

The results should also have important clinical implications. Most current treatments were discovered by serendipity, rather than based on an understanding of the causes or mechanisms of disease. Understanding which genes are involved in a disorder and how they relate to brain function at the molecular level should reveal likely targets for more effective medicines.

The **Charles Nemeroff** article on the two major options to treat depression emphasizes how difficult it has been to develop new medications without greater understanding of the most appropriate molecular targets. He also articulates how difficult it is for a clinician to decide which among similar but not identical medications would be most effective for individual patients. Not all patients respond in the same way to the same medication; some do better on one, while others do better on another. Nemeroff uses depression to illustrate the problem: More than 30 medications are approved by the Food & Drug Administration for the treatment of depression, and many of them work in quite similar ways. How does a clinician know which medication to prescribe for which patient? For now, a patient's past treatment response seems the best predictor of which medications to try with a recurrent illness, though that often does not work. Nemeroff discusses the probability that, in time, specific metabolic parameters might be used to help do a better job of deciding which medications to try, and whether to try them with or without the simultaneous use of different behavioral therapies.

Understanding individual differences is critical to predicting or understanding which patients are more or less vulnerable to diverse conditions. **George Koob** outlines the mechanisms of stress responses and points out how individual differences in reactivity to stress can predispose different individuals to negative emotional states. His article proposes that individual differences in stress vulnerability and resilience, which are key elements of the development of post-traumatic stress disorder and addiction, derive from differences in activation of the neurocircuitry of negative emotions. He argues that "new advances in our understanding of the neurocircuitry of the dark side and identification of epigenetic factors that weigh the function of these circuits will be the key to precision medicine for the diagnosis and treatment of these disorders."

Reading about oxytocin and the hypothalamo-pituitary-adrenal hormones and their relationship to behavior in the **George Koob** and **Paul Zak** articles brought to mind an example from my own career where today's scientific heresy can become tomorrow's fact. Some 30 years ago, my research program focused on the role of neuropeptides and pituitary adrenal hormones in social behavior, primarily using animal models. Among our discoveries was that a short-term injection regimen of the adrenal steroid corticosterone leads to an increase in inter-male aggressiveness in mice. No one had shown behavioral effects of these steroids before, and the reaction of a National Institutes of Health review committee was that I should consider another line of research, since it was impossible for those hormones to have direct behavioral effects. I did persist, though, and reapplied with aslightly different argument, only to be met with an even stronger message that this line of research could not possibly go anywhere. I moved on to other things, but many years later, a young scientist came up to me at a conference to tell me how excited she was to replicate those findings, and it is now well known clinically, of course, that particularly long-term treatment with synthetic adrenal steroids, like dexamethasone, often leads to increased irritability in diverse groups of patients. Authors of papers studying oxytocin effects on pair bonding in rodents were met with similar reactions, only to have their findings replicated and extended much later, in much the same way.

Multiple papers deal with compulsive behaviors, including the one by **Alice Ely** and **Anne Cusack** on binge eating, the one by **Theodore Cicero** on the abuse of prescription narcotics, and the piece by **Robin Murray** about the risks of heavy marijuana use. They discuss well the great progress being made on each of these problems. They also collectively illustrate how our understanding of the nature of addictive and compulsive behaviors has evolved, and, in the cases of prescription narcotic and marijuana use, how they are health issues as well as criminal justice issues. These papers also emphasize that there are biological commonalities, as well as idiosyncrasies, to the diverse compulsions and addictions to which humans fall prey.

Once again, the *Cerebrum* anthology provides a very interesting and provocative collection of articles illustrating the diversity of modern neuroscience research and its broad implications for the human condition. The papers are clear and written at a level enabling all kinds of readers to find them understandable and very useful.

ARTICLES

1

Appraising the Risks of Reefer Madness

By Sir Robin Murray, Ph.D.

Sir Robin Murray, Ph.D., is professor of psychiatric research at the Institute of Psychiatry, Kings College, London. He has contributed to the understanding that schizophrenia is more than a genetic brain disease; environmental factors such as obstetric events, cannabis abuse, migration, and adverse life events also increase the risk of the disorder. He is involved in testing new treatments for schizophrenia, and cares for people with psychosis at the Maudsley Hospital in London. Murray has written over 700 articles and is one of only five psychiatrists ever to have been elected a Fellow of the UK Royal Society (Freud was the first). A former president of the European Psychiatric Association and co-editor-in-chief of *Psychological Medicine*, he received a knighthood from the Queen in 2011.

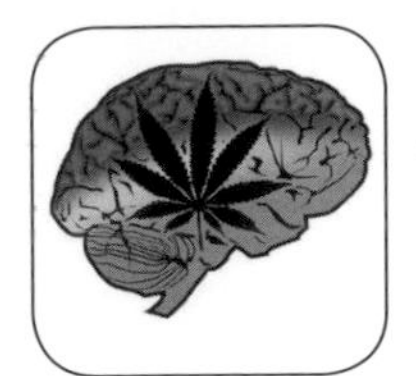

Editor's Note: Studies that have tied cannabis use to schizophrenia in the developing brain are just the tip of the iceberg when it comes to marijuana. Are different strains and synthetic cannabinoids especially dangerous? Are we doing enough to educate young people on the risks? Does marijuana use lower IQ? Where is the line between medical marijuana and recreational use? Our author, a noted British psychiatrist, offers a European perspective on these issues.

BEGINNING IN THE MID-1980s, European psychiatrists like me started seeing an increasing number of previously well-functioning teenagers who had developed hallucinations and delusions: the characteristic picture of schizophrenia. These troubled patients puzzled us because most had been bright and sociable and had no ties to the usual risk factors such as a family history of the disorder or developmental insult to the brain. Family and friends would often say, "Maybe it was all the cannabis they have been smoking," and we would confidently reassure them that they were mistaken and tell them that cannabis was known to be a safe drug.

My view began to shift when a colleague, Peter Allebeck from the Karolinska Institute in Stockholm, launched his own investigation. He had been struck similarly by seeing well-adjusted young people develop schizophrenia for no apparent reason. The wonderful Swedish national records system enabled him to trace the outcome of 45,750 young men who had been asked about their drug use when they were conscripted into the Swedish army. From analysis of these data, Allebeck and his colleagues[1] reported in 1987 that conscripts who had used cannabis more than 50 times were six times more likely to develop schizophrenia over the next 15 years than those who had never used it (Figure 1).

Because only one study existed, most psychiatrists reassured themselves that the cannabis users probably were destined for schizophrenia before they started smoking. As a result, Allebeck's findings were mostly ignored. Even the prestigious medical journal *The Lancet*, which had published Allebeck's paper, carried an editorial in 1995 that restated the prevailing medical view that "the smoking of cannabis, even long term, is not harmful to health."[2]

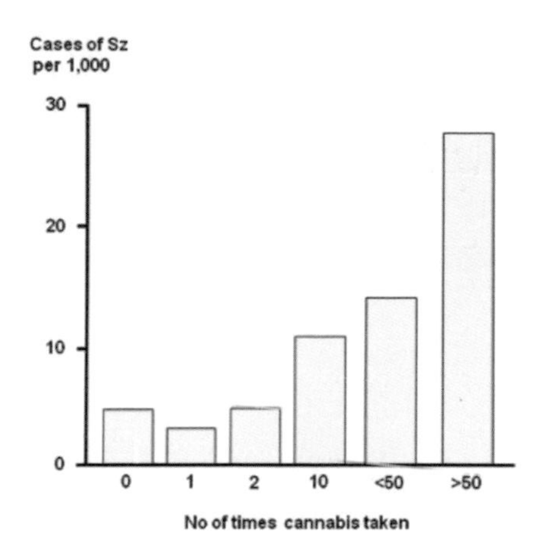

Figure 1. Risk of schizophrenia in conscripts followed up for 15 years

Nevertheless, with the help of a young psychiatrist, Anton Grech, I began looking at patients who had smoked cannabis and been admitted to the Maudsley Hospital in London with a diagnosis of psychosis. Our aim was to see what happened to the ones who continued to use cannabis. If, as some suggested, psychotic patients used cannabis to "self-medicate" or otherwise help them to cope with their illness, then one might expect the persistent cannabis users to have a better outcome. But we found the opposite; four years later, the patients who continued to use cannabis were much more likely to still have delusions and hallucinations.[3] Many studies have replicated the findings.[4-6]

So, if smoking cannabis exacerbates established psychosis, might it play a causal role too? This question provided the impetus for various research groups to try to replicate the Swedish army study by following- up samples of young people in the general population, divided into cannabis users and non-users. Since 2002, a series of ten such studies have reported that individuals who used cannabis at the baseline evaluation had a greater risk of subsequently developing psychotic symptoms and indeed full-blown schizophrenia than non-users.[7-18] Other studies of cannabis users who had sought medical care showed that they had a marked increased risk of subsequent schizophrenia.[19-20] Such unanimity is rare in psychiatric epidemiology.[21]

A Hostile Reaction

Much of the scientific establishment greeted these findings with hostility, as did people advocating a more liberal attitude regarding cannabis use. My colleagues and I were accused of being closet prohibitionists; some suggested we were experiencing our own kind of "reefer madness." Critics suggested that people using cannabis were doing so because they were odd and destined to develop schizophrenia anyway.[22]

But in a study of young people who had been scrutinized intensely from birth in Dunedin, New Zealand, we were able to exclude those who already appeared psychosis-prone at age 11. We still found a link between cannabis use and later schizophrenia, even when we excluded the effects of other drugs known to increase risk of psychosis.[7] Another criticism was that maybe some people were taking cannabis in an attempt to ameliorate symptoms of psychosis or its precursors. However, a second New Zealand study, this time from Christchurch, showed that once minor psychotic symptoms developed, people tended to smoke less.[8]

Eventually the sheer volume of data convinced European and Australian psychiatrists of a link. Cannabis is now generally accepted as a cause of schizophrenia[23-24] (though less so in North America, where this topic has received little attention). Argument does continue over just how significant cannabis-associated psychosis is. In different countries, the proportion of schizophrenia attributed to cannabis use ranges from 8 to 24 percent, depending, in part, on the prevalence of cannabis use.[25]

Excess Becomes a Major Concern

Most people enjoy cannabis in moderation and suffer few or no adverse effects. People take comfort in knowing that some of the most celebrated role models on the planet—Paul McCartney, Oprah Winfrey, and Barack Obama, for example—have admitted to smoking cannabis with no ill effect. But the scientific consensus is that, although there is no convincing evidence of a link to anxiety[26] or depression,[27] cannabis can lead to psychosis when consumed in large amounts. The bottom line: daily users who smoke large amounts are increasing their risk of schizophrenia.

Nevertheless, the vast majority of users won't become psychotic. Indeed, when young people who have developed schizophrenia after years of smoking cannabis are asked whether they think their habit may have contributed, they often say, "No, my friends smoke as much as I do, and they're fine." It appears that some people are especially vulnerable.

Not surprisingly, people with a paranoid or "psychosis-prone" personality are at greatest risk, alongside people with a family history of psychosis.[28] Research also suggests that inheriting certain variants of genes that influence the dopamine system, which is implicated in psychosis, may make some users especially susceptible; examples of such genes include AKT1, DRD2, and possibly COMT.[29-31]

Another important question is: Are some types of cannabis more risky than others?

The Changing Nature of Cannabis

The cannabis plant is thought to originate from either central Asia or the foothills of the Himalayas. It produces compounds known as cannabinoids in glandular trichomes, mostly around the flowering tops of the plant. Recreational cannabis is derived from trichomes and has been traditionally available as herb (marijuana, grass, weed) or resin (hashish, hash); the former is most common in North America and the latter in Europe. Cannabis acts on the CB1 cannabinoid receptor, part of the endocannabinoid system, which helps to maintain neurochemical stability in the brain. The cannabis plant produces more than 70 cannabinoids, but the most important are tetrahydrocannabinol (THC) and cannabidiol (CBD). THC is responsible for the "high" that users enjoy. It activates the cannabinoid CB1 receptor, which is one of the most widespread receptors in the brain.

The proportion of THC in traditional marijuana and resin in the 1960s was approximately 1 to 3 percent. The potency of cannabis began to rise in the 1980s, when cannabis growers such as David Watson, commonly known as "Sam the Skunkman," fled the Reagan-inspired "War on Drugs" and brought cannabis seeds to Amsterdam, where it could be sold legally in "coffee shops." Together with Dutch enthusiasts, they experimented to produce more potent plants. This set the trend for a slow but steady increase in a new variety of marijuana called sinsemilla, harvested from unpollinated female flowers (often called "skunk" because of its strong smell). By the early years of the 21st century in England and Holland, the potency of sinsemilla, as measured by the proportion of THC, had risen to between 16 and 20 percent, respectively, and it had taken over much of the traditional market from resin. [32-33]

In 1845, French psychiatrist Jacques-Joseph Moreau (nicknamed "Moreau de Tours") took cannabis himself in the appropriately named Club de Hashischins and gave it to some of his students and patients.[34] He concluded that cannabis could precipitate "acute psychotic reactions, generally lasting but a few hours, but occasionally as long as a week." However, since then surprisingly little formal scientific research has explored THC's effects in humans.

One of the first THC studies was conducted by Paul Morrison, a Scottish psychopharmacologist in our group at King's College London, who gave the ingredient intravenously to healthy and eager young volunteers.[35] Their reactions ranged from euphoria to suspicion, paranoia, and hallucinations. Later, he and Amir Englund[36] pretreated their volunteers with CBD, the other main ingredient of traditional cannabis, before giving them intravenous THC; the effects of the latter were much diminished. Thus, CBD appeared to counter the psychotogenic effects of THC. A German clinical trial supports this idea: Marcus Leweke[37] found that CBD had antipsychotic actions equivalent to a standard antipsychotic, amisulpride, in patients with schizophrenia.

High-potency types of cannabis such as sinsemilla (skunk) differ from traditional forms not only in the amount of THC they contain but also in the proportion of CBD. Interestingly, plants bred to produce a high concentration of THC cannot also produce a lot of CBD, so the high-THC types of cannabis contain little or no CBD.[32]

Might high-potency types of cannabis be more likely, therefore, to induce psychosis than traditional forms? To examine this question, my wife, Marta Di Forti, a psychiatrist supported by the UK Medical Research

Council, compared the cannabis habits of 410 patients who were admitted to the Maudsley Hospital with their first episode of psychotic disorder to those of 390 controls in the local population. Those who had been using high-potency cannabis (skunk) had a much higher risk of psychosis than users of resin.[25,38-39] People using skunk-like cannabis on a daily basis were five times more likely than non-users to suffer from a psychotic disorder, while users of traditional resin did not differ from non-users. Another study which tested hair for cannabinoids, showed that those users with both detectable THC and CBD in their hair had fewer psychotic symptoms than those with only THC.[40]

Thus, the increasing availability of high-potency types of cannabis explains why psychiatrists should be more concerned about cannabis now than they were in the 1960s and 1970s. The trend toward greater potency is not slackening, with new forms of resin and resin oil being reported to contain up to 60 percent.[41] These particular very potent forms remain unusual, but synthetic cannabinoids (often termed "spice" or "K2") are now commonly advertised and sold on websites that keep within the law by labeling their products as incense—or adding "not for human consumption." While THC only partially activates the CB1 receptor, most spice/K2 molecules fully activate the receptor and therefore are more powerful. Consequently, acute adverse reactions are more common. A survey of 80,000 drug users showed that those who used synthetic cannabinoids were 30 times more likely to end up in an emergency room than users of traditional cannabis.[42] The agitation, anxiety, paranoia, and psychosis that can result from synthetic cannabinoids use have been dubbed "spiceophrenia."[43-44]

Dependence and Cognitive Impairment Remain Controversial

While psychological dependence and tolerance occur, the issue of physical dependence remains hotly debated. Advocates point to an apparent absence of acute withdrawal symptoms from cannabis (there have been no studies to date examining specifically high-THC users). Under ordinary circumstances, withdrawal symptoms are absent because cannabis remains in the body for several weeks—so withdrawal is very gradual and not obvious, though anxiety, insomnia, appetite disturbance, and depression can develop. Some claim that 10 percent of people who experiment with cannabis will develop dependence and that rates of dependence among daily users are at least 25 percent. Whether or not such figures are true, cannabis dependence is an increasingly common cause of those seeking help in Australia, Europe, and North America.[45-47]

Another controversial issue is cognitive impairment. THC disrupts the hippocampus, the area of the brain crucial to memory. When Paul Morrison induced psychotic symptoms by giving intravenous THC to volunteers, transient cognitive impairment also emerged. Such impairment likely is why drivers under the influence of cannabis are at double the risk of traffic accidents.[47]

Heavy users show cognitive impairment, but disagreement continues over what happens when they stop. using. Some studies suggest they can recover fully, while others indicate that only partial recovery is possible.[48] One (as yet unreplicated) study that provoked enormous concern was a report by Madeleine Meier and her colleagues, based on the Dunedin study.[49] This suggested that persistent cannabis use over several decades causes a decline of up to eight points in IQ.

One possible explanation for the inconsistent cognitive findings is that the effects on cognition might depend on the age at which cannabis use began. Harrison Pope and colleagues examined long-term heavy cannabis users and found that it was those who initiated cannabis use before age 17 who showed lower verbal IQ scores.[50] Meier also found greater decline in those in her Dunedin cohort who started using in adolescence.[49]

Other studies have implicated an association between adolescence cannabis use and poor educational achievement. Edmund Silins and colleagues reviewed more than 2,500 young people in Australia and found that daily cannabis use before age 17 was associated with "clear reductions" in the likelihood of completing high school and obtaining a university degree.[51]

Researchers have raised similar questions regarding age of initiation and risk of psychosis. In our Dunedin study, we found that people who started to use cannabis at age 18 or later showed only a small, non-significant increase in the risk of schizophrenia-like psychosis by age 26. But among those starting at age 15 or earlier, the risk increased fourfold[7] (Figure 2). Other studies have reported similar disparities.[21]

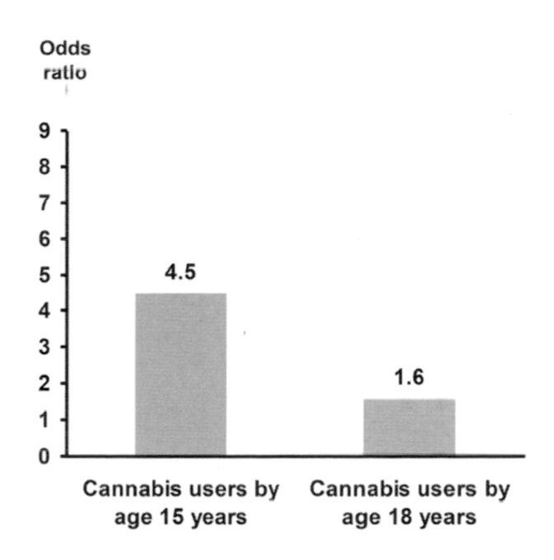

Figure 2. Risk of schizophrenia-like psychosis at age 26 years

Experimental studies in rodents have revealed that THC administration produced a greater impact on cognitive function in juvenile rats than in adult rats.[21] Also, some recent brain imaging studies of long-term, very heavy cannabis-using people have claimed to find detectible brain changes, especially in those who started in adolescence.[52-54] Although the imaging studies remain contentious, a possible explanation is that beginning cannabis use at an age when the brain is still developing might permanently impair the endocannabinoid system and impact other neurotransmitters such as the dopamine system[21]—known to be implicated in both learning and in psychosis.

Implications

Cannabis attitudes are changing globally. Uruguay has legalized its use, as have four American states. In addition, 17 states have decriminalized cannabis, while 23 others have passed "medical marijuana" laws on the basis that it helps people with chronic pain and symptoms associated with chemotherapy. A curious divide has opened up between North America and Europe. In the U.S., cannabis use in young people has increased since the mid-1990s as the number regarding use of cannabis as risky has fallen.[55] In contrast, use has fallen in many European countries, partly because of greater knowledge of the risks to mental health. In England for example, in 1998, 26 percent of people ages 16 to 24 admitted to having used cannabis in the previous year. By 2013, that number had declined to 15 percent.[56]

As they consider the future of legal cannabis use, legislators and politicians need to balance the enjoyment of the many with the potential for harm. That challenge is displayed in the United Kingdom, where pro- and anti-cannabis lobbies have engaged in a raucous argument. In 2004, the government took the advice of the Advisory Council on the Misuse of Drugs that the use of cannabis had no serious adverse effects, and lowered restrictions on the drug. The following year, the Advisory Council reversed its view and accepted the evidence that linked cannabis to psychosis. Some newspapers that had campaigned for liberalization responded by turning on the "liberal" politicians. By 2007, the media clamor had reached such a pitch that the government increased restrictions again. Now pressure is building up once again for liberalization.[57]

In picking their way through the conflicting views, politicians and regulators need to recognize that "medicinal marijuana" has become largely a cover for introducing recreational use by the marijuana industry,[58] and that unscrupulous and increasingly wealthy doctors are involved. However, research into the numerous components of cannabis should be encouraged since, like research into opiates, it may produce drugs with important medical uses. Individual cannabinoid components can then be subject to trials measuring their effectiveness for a variety of ailments (e.g. multiple sclerosis, epilepsy) in the same way any other proposed drug is evaluated. When effective, it should be introduced for prescription by doctors; several cannabinoid drugs already have become available in this manner.

But we must be careful, since the evidence we have reviewed suggests that cannabis, like other psychotropic drugs, has negative as well as positive effects. When you consider that almost 200 million people worldwide use cannabis,[59] the number of people who suffer cannabis-induced psychosis is likely to be in the millions, and the number at risk for developing serious mental health problems becomes a huge concern.

Europeans are watching the evolution of the cannabis debate in the U.S. with great interest. Decriminalization and legalization of cannabis and synthetic cannabinoids are all on the table. Will legalization mean an increase in consumption? Will this result in greater use by those in their early teens, who seem most susceptible to adverse effects? Will the mental health and addiction services be able to cope? How will educational campaigns regarding the risks of regular use of high-potency cannabis or synthetic cannabinoids play out? Might a simple genetic test reveal who is likely to suffer adverse mental effects? As cannabis use continues to win acceptance, as many questions as answers remain.

2

Why Inspiring Stories Make Us React

The Neuroscience of Narrative

By Paul J. Zak, Ph.D.

Paul J. Zak, Ph.D., is a scientist, author, and public speaker. His book *The Moral Molecule: The Source of Love and Prosperity* was published in 2012 and was a finalist for the Wellcome Trust Book Prize. He is the founding director of the Center for Neuroeconomics Studies and professor of economics, psychology, and management at Claremont Graduate University. Zak also serves as professor of neurology at Loma Linda University Medical Center. He has degrees in mathematics and economics from San Diego State University, a Ph.D. in economics from University of Pennsylvania, and postdoctoral training in neuroimaging from Harvard. He is credited with the first published use of the term "neuroeconomics" and has been a vanguard in this new discipline. He organized and administers the first doctoral program in neuroeconomics. Zak's lab discovered in 2004 that the brain chemical oxytocin allows us to determine who to trust. His current research has shown that oxytocin is responsible for virtuous behaviors, working as the brain's "moral molecule."

Editor's Note: The man behind the discovery of the behavioral effect of a neurochemical in the brain called oxytocin wondered if the molecule might motivate people to engage in cooperative behaviors. In a series of tests using videos, his lab discovered that compelling narratives cause oxytocin release and have the power to affect our attitudes, beliefs, and behaviors.

DURING A NIGHT FLIGHT HOME to California after five days in Washington, D.C., I discovered that I am the last person you would want sitting next to you on a plane. Tired and unable to bang on my laptop in the turbulence at 40,000 feet, I decided to watch *Million Dollar Baby*. I hadn't seen it, but I figured a Clint Eastwood–directed film that had won the Oscar for Best Picture would be a deserved break for a hard week.

It is a wonderful film, and I became deeply absorbed in it. The narrative is circumscribed by a father-daughter story and concludes with an agonizing act. When the movie was over, the man next to me said, "Sir, is there something I can do to help you?" I was crying. Well, not really crying, more like heaving big sloppy sobs out of my eyes and nose and mouth. Everyone around could hear me but I could not suppress my sadness.

After I recovered, I began to wonder what had happened to me. I was cognitively intact, aware of my surroundings and who I was. And yet the story was so engaging that it caused my brain to react as if I were a character in the movie, as if one of my own daughters were the one suffering. I experienced heartache as the movie ended, but then it was only a story.

As a neuroscientist, I knew that movies changed our brain activity in some way, but how?

I soon realized I had stumbled on a potentially useful way to extend my studies of the social brain. My lab was the first to discover that the neurochemical oxytocin is synthesized in the human brain when one is trusted and that the molecule motivates reciprocation.[1,2,3] We found that the human oxytocin response was similar to that found in social rodents,[4] signaling that another person (or rodent) is safe and familiar. Perhaps most surprising, we found that in humans, this "you seem trustworthy" signal occurs even between strangers without face-to-face interactions.

Oxytocin is an astonishingly interesting molecule. It is a small peptide synthesized in the hypothalamus of mammal brains. It is made of only nine amino acids and is fragile. Oxytocin is classically associated with uterine contractions and milk-letdown for nursing. Animal studies have shown that under physiologic stress oxytocin is released in both brain and body.[5,6] This is unusual for a brain-derived neurochemical, but it provides a powerful way to study oxytocin: After a stimulus, changes in oxytocin in blood reflect changes in the brain's oxytocin.

For more than a decade, I have run human experiments measuring the endogenous release of oxytocin during social interactions. My colleagues and I have studied oxytocin release in the laboratory as well as in field studies spanning religious rituals, folk dances, weddings, and a traditional war dance by indigenous people in the rainforest of Papua New Guinea.[7] I also have demonstrated the causal effect of oxytocin on prosocial behaviors by safely infusing synthetic oxytocin into hundreds of people's brains through their noses.[8,9,10] Oxytocin infusion increases prosocial behaviors. It's like turning on a garden hose and watching the water spray out.

Provoking the Brain

Studies that only infuse oxytocin into participants and then make claims about human behavior are suspect. This approach does not identify what the brain itself is doing during social interactions, including neurochemical promotion and inhibition of oxytocin synthesis, and dose-response relationships between oxytocin and behavior. The key question is whether the brain produces its own oxytocin during the behavior being studied; if so, the causal relationship between oxytocin and a particular behavior can be demonstrated via an infusion study. But the reverse is not true: Infusing oxytocin or any drug into the brain and observing a change in behavior does not mean that this is how the brain works—it simply means that a drug has changed behavior, as many drugs do. My studies complete the causal circle by measuring what the brain does naturally and then intervening in this system pharmacologically to show that the behavior can be provoked.

After years of experiments, I now consider oxytocin the neurologic substrate for the golden rule: If you treat me well, in most cases my brain will synthesize oxytocin and this will motivate me to treat you well in return. This is how social creatures such as humans maintain themselves as part of social groups—they play nice most of the time. (Why people do not play nice is a fascinating story we also have studied; see Zak, 2012, for evidence). But I'm a skeptic at heart, so I always want to measure the behavioral effects of oxytocin rather than simply ask people's opinions about how they feel.

The experience I had watching *Million Dollar Baby* caused me to wonder if movies, in addition to direct personal interactions, would cause oxytocin release. To test this, my colleague Jorge Barraza edited a set of a short video clips that we obtained with permission from St. Jude Children's Research Hospital. One version shows a father talking to the camera while his two-year-old son, Ben, who has terminal brain cancer, plays in the background. The story has a classic dramatic arc in which the father is struggling to connect to and enjoy his son, all the while knowing that the child has only a few months to live. The clip concludes with the father finding the strength to stay emotionally close to his son "until he takes his last breath."

We also developed a video of the same father and son spending a day at the zoo. This version does not mention cancer or death, but the boy is bald (from his chemotherapy) and is called "miracle boy" once during the clip. This video lacks the tension induced by the typical story form but includes the same characters. This version was used as a control story to see what the brain does when any video is being watched.

In our first study of narratives, we took blood before and after participants watched one of the two versions of the video.[11] We found that the narrative with the dramatic arc caused an increase in cortisol and oxytocin. Tellingly, the change in oxytocin had a positive correlation with participants' feeling of empathy for Ben and his father. Heightened empathy motivated participants to offer money to a stranger who was in the experiment. We connected a story to a feeling and then to a prosocial behavior. The "flat" narrative of Ben and his father at the zoo did not increase oxytocin or cortisol, and participants did not report empathy for the story's characters.

These findings suggest that emotionally engaging narratives inspire post-narrative actions—in this case, sending money to a stranger. But maybe this result only applied to videos of dying children. Also, we did not know for sure that oxytocin was the reason participants cared about the people in the video, just that oxytocin and empathy were correlated. So we rolled up our sleeves and ran more experiments.

Narrative Immersion

Our previous study pointed to oxytocin as the biological instrument that puts people in thrall to a story. To assess the causal impact of oxytocin on narrative immersion, we ran a study using public service announcements (PSAs) in which participants received intranasal infusions of synthetic oxytocin or a placebo. This time around, we decided to test a larger set of video narratives. We wanted stories that most people would not have seen before and ones that could elicit a prosocial action at a cost (such as a donation). This would allow us to measure objectively whether the story "got to you."

We found a rich trove of public service announcements from the United Kingdom that are well-produced and engaging. The experiment used 16 PSAs that ran for 30 or 60 seconds on four topics: smoking, drinking to excess, speeding, and global warming. To incentivize people to pay attention to the videos, each of the participants was paid five dollars if they could correctly answer a factual question about the ad immediately after watching it. For example, "Was there a car in the video?" Then, our software asked participants if they would like to donate some of the five dollars they had just earned to a charity associated with the cause shown in the PSA. None of the PSAs solicited donations, they simply told stories about social issues. Computer software presented all the videos and post-video questions, and we used random participant identifiers so that one's donation behavior was kept private.

Forty people received either 40 units of oxytocin or an equivalent amount of normal saline (placebo). Neither the experimenters nor the participants knew what substance had been administered. Participants started watching the videos after an hour-long period during which the synthetic oxytocin diffuses from the sinuses into the brain.

We found that those who received oxytocin donated, on average, 56 percent more money to charity compared with participants who received the placebo.[12] This confirmed the causal role of oxytocin on post-narrative prosocial behavior. But why did this happen? We discovered that participants who were given oxytocin showed substantially more concern for the characters in the PSAs. This increased concern motivated them to want to help by donating money to a charity that could alleviate the suffering these stories depicted.

If you think about it, the donations are quite odd. The narrative is over, but the effects linger. It is as if the brain is lazy and is using a "monkey see, monkey do" approach to assess appropriate social behaviors. (Indeed, the brain seeks to conserve energy by using default pathways—a kind of "laziness.") The PSAs seemed to persuade viewers that (for example) nowadays the humans are very concerned about drinking too much, so as a human, I, too, should be concerned. And I should demonstrate that concern by donating money to charity. Such responses are what social creatures with social brains do. And yet, participants understand that the stories are fictional and are portrayed by professional actors. The money donated to charity cannot help these actors out of their fictional binds. The money might help prevent the harm depicted in the PSAs from happening to an unknown other person, but this is a big "if." Nevertheless, oxytocin makes people want to help others in costly and tangible ways.

In another experiment,[12] we sought to replicate our earlier study by taking blood samples before a group of 42 participants (who were not in the oxytocin infusion study) watched one of the UK PSAs. We measured the change in oxytocin and in a fast-acting arousal hormone with a long name that is abbreviated ACTH.

When the PSA elicited an increase in both ACTH and oxytocin, donations were 261 percent higher than when one or both of these biomarkers did not rise. The change in ACTH correlated with the amount of attention people paid to the story. This finding makes sense: If we do not attend to a story, it will not pull us into its narrative arc. Attention is a scarce neural resource because it is metabolically costly to a brain that needs to conserve resources. If a story does not sustain our attention, then the brain will look for something else more interesting to do.

We also found that the change in oxytocin was associated with concern for the characters in the story, replicating our earlier finding. If you pay attention to the story and become emotionally engaged with the story's characters, then it is as if you have been transported into the story's world. This is why your palms sweat when James Bond dodges bullets. And why you stifle a sniffle when Bambi's mother dies.

Attention-Getting

Narratives that cause us to pay attention and also involve us emotionally are the stories that move us to action. This is what a good documentary film does. More generally, stories with a dramatic arc fit the requirements for high-impact narratives. This structure sustains attention by building suspense while at the same time providing a vehicle for character development. The climax of the story keeps us on the edge of our neural seats until the tension is relieved at the finish.

Theorists including Aristotle (*Poetics*, 335 BCE), Gustav Freytag (*Die Technik des Dramas*, 1863), and Joseph Campbell (*The Hero with a Thousand Faces*, 1949) have contended that the rising and falling tension of dramatic performances facilitate the audience's emotional connection to the characters. Hollywood writers call this creating "surprising familiarity." Every story is different but somehow the same.

Now let's get down to brass tacks: Why are there so many dreadful movies? Humans have known about the three-act structure and mythos, pathos, and ethos for 2,500 years. This is where the neuroscience hits the flickering screen.

Like all experiments, we had to start small.

To answer these questions, we needed to measure attention and oxytocin responses rapidly—second by second, or even faster. Blood draws would not do. At the same time, the U.S. Department of Defense wanted to know why narratives are persuasive, and supported our research and that of other labs as well. Attention is easy to measure rapidly, via a quickened heartbeat or sweat coming from eccrine glands in the skin. But was there a way to measure oxytocin rapidly? Nature provided a solution. While we were mostly interested in oxytocin in the brain, the stimulus-induced co-release of oxytocin in the brain and blood meant we could measure changing activity in regions with densities of oxytocin receptors. The vagus nerve (the longest cranial nerve, which innervates the heart and gut) is chock-full of oxytocin receptors. With a bit of algorithmic fiddling, scientists can measure the activity of the vagus using an electrocardiogram (ECG). We confirmed that the change in oxytocin in blood correlates with changes in vagus nerve activity. Voilà, we had a measurement technique. But would it predict behavior?

We returned to the story of the dying child Ben because it is a reliable way to stimulate oxytocin release. This time we measured cardiac activity using an ECG and sweat using an electrodermal sensor on the fingers. Because we were developing a system that might be used in a war zone, we built in redundancies. Attention was measured using both heart rate and skin conductance changes from sweat on the fingers; emotional resonance was quantified using two measures of changes in the brain's relaxation response driven by the vagus nerve. The exciting part was that we could measure both effects up to 1,000 times a second with off-the-shelf wireless technologies.

But it is not so simple to isolate the effects of a story from everything else the brain is doing to keep you upright, breathing, and conscious. All neuroscience studies need to extract the neurologic signal produced by a stimulus during an experiment from the background noise of all other neural activity. To give you a sense of the scope of this problem, for every 30 people we test for an hour each, we collect a terabyte of peripheral neurologic data. Most of this data is not relevant to understanding why people respond to stories, but the faint traces that are relevant must be extracted and processed with extraordinary care. Once we did all this, the data told us several interesting things.[13]

First among them is that the brain does not work like the hypothetical story structure known as Freytag's pyramid, in which strictly rising action leads to a climax and then strictly falling action occurs as the story resolves. Even for the 100-second "Ben" video, one's attention waxes and wanes. The brain is attending to the story and then doing a quick search of the rest of the environment, and then refocusing on the story as the tension rises. Nevertheless, the peak attentional response occurs in the climax, when Ben's father reveals that Ben is dying. That's a bombshell to which people pay attention.

The oxytocin response lags behind the attentional spike as the story begins. After about 30 seconds, vagal activity begins to increase as viewers get to know and then begin to empathize with Ben and his father. Attention to the story provides a reason viewers should care about the characters.

Not only were we able we track what the brain is doing millisecond by millisecond during a story, we used the neurologic data to build a predictive model of donations to a childhood cancer charity—our measure of story impact. The statistical model we built predicts whether a participant would donate money, with 82 percent accuracy. That is, by measuring how your peripheral nervous system responds to a story, we can almost perfectly predict what you'll do before you do it.

The participants who, for whatever reason, either lost interest in the video or didn't form an emotional connection to Ben and his father almost never donated money to charity. But we are still left with a mystery: Why donate money at all? The money will not save Ben and it won't offer relief to his father. It seems that once we are attentive and emotionally engaged, our brains go into mimic mode and mirror the behaviors that the characters in the story are doing or might do. As social creatures we are biased toward engaging with others, and effective stories motivate us to help others.

Truth be told, Ben's story is as near to a perfect high-impact narrative as there is. We wondered if neurologic data could identify bad stories, too. And what about stories that may be distasteful but that are still desirable to watch? I watched Steven Spielberg's Holocaust movie *Schindler's List* once. I'm glad I did, but I don't have much desire to watch it again. It was just overwhelming emotionally.

Our next study tested stories about "hot-button" issues to see how people reacted to potentially disagreeable topics. We used first-person narratives from StoryCorps, a nonprofit that collects and distributes personal stories. We choose six stories on racism, gun control, and the terrorist attacks of September 11. Each anecdote lasted from two to four minutes. For our "narrative impact" measure, we invited participants to donate some of their earnings to a charity associated with the topic of the story.

These stories were challenging to analyze because they varied substantially in structure and content. The peripheral neurologic data we collected reflected these variations. Just as in the "Ben" story, we confirmed that stories that sustain attention and generate emotional resonance produce post-narrative donations—even stories on difficult topics. To the brain, good stories are good stories, whether first-person or third-person, on topics happy or sad, as long as they get us to care about their characters.

Psycholinguists have shown that effective stories induce "transportation" into the narrative.[14] Transportation happens when one loses oneself in the flow of the story—just like I did while watching *Million Dollar Baby.* To understand the psychological effects of stories, we included surveys of narrative transportation and concern for story characters in the StoryCorps study. Both narrative transportation and concern predicted post-story donations. This shows why stories affect behavior after the story has ended: We have put ourselves into the narrative. Even a week after the experiment, accurate story recall was predicted by a single measure: narrative transportation.

Do We Know a Good Story When We See One?

You may be thinking that we have a money-centric approach to assessing when people are moved by a story. Fair enough. Let's try a different approach: We'll have thousands of people rate stories instead. The stories we used were TV commercials. Conveniently, this is just what *USA Today* asks readers to do on Super Bowl Sunday: vote for the commercials they like the best. About 5,000 people voted for their favorite commercials in 2014, and the style and content of these short narratives vary from the unusual to poignant to just plain silly. This gave us a chance to further refine our algorithms and test them against what people say they like.

USA Today does not simply provide a ranking of commercials; it has its readers rate them on a one-to-ten scale. Good idea! My group derived a quantification of narrative engagement using neurologic data so we, too, could rate story quality. We estimated the relative contribution of attention and emotional resonance on story impact from our corpus of studied stories. We call this measure a story's ZEST (for Zak Engagement STatistic). By estimating each Super Bowl ad's ZEST, we could compare the *USA Today* readers' ad likability with the ZEST measure of brain activity.

Three days after the 2014 Super Bowl, 16 participants watched the top ten Super Bowl commercials in random order in my lab while we measured their peripheral neurologic activity. The results were astounding. There was no correlation at all between what *USA Today* readers said they liked and a commercial's ZEST. Either we had made a big mistake, or we had discovered something important. So we ran another study using *USA Today*'s top ten 2013 Super Bowl commercials and found exactly the same thing: zero correlation.

These findings suggest that people are unable to articulate what they like and do not like. But their brains reveal what is engaging for them to watch. Perhaps this should not surprise us. In a classic study, psychologist and economics Nobel laureate Daniel Kahneman found that people's preferences for things they have not experienced are largely unformed.[15]

Watching the Super Bowl commercials myself, I sensed why it is hard to articulate what one likes. The best Super Bowl commercial in 2014, according to *USA Today* readers, was called "Puppy Love," produced for Budweiser beer. In the first ten seconds, one sees a puppy nuzzling the nose of a Clydesdale horse. One immediately recognizes the Clydesdale as the Budweiser icon, and this tells viewers what they can expect from the ad. The suspense is gone, and our neurologic measures show that people's attention wanders starting 15 seconds into the commercial. Without attention, the hoped-for emotional resonance with the ad's characters (and presumably the brand) fails to occur.

But ask people what they like and, gosh, they see puppies and horses and wide open country and, well, of course we love these images. But the brain does not lie. The commercial is dull.

In all our studies we ruled out effects that might influence ZEST, including movement, cars, buildings, attractive men and women, and many other factors. They don't matter; it all comes down to story.

The U.S. Department of Defense's funding of the emerging science of narrative jump-started the field.[16,17] Storytellers have always known that attention and emotion are important to develop during a narrative, but now we have ways to measure these responses directly rather than rely on inchoate impressions such as "entertaining" or "fascinating." Yet, even with millennia of practice, creating a great story is difficult. The emerging science of narrative can guide the art, but it cannot replace it. Humans are just too complex for an algorithm to generate art. And this is where the artist comes in. The narrator in *Million Dollar Baby* describes the heroine, Maggie's, desire to be a boxer as "...the magic of risking everything for a dream that nobody sees but you." Artists who create worlds we cannot help but enter do the same.

3

Vernon Remembered

Editor's Note: The world of neuroscience lost one of its pioneers when Vernon B. Mountcastle, M.D., died January 11 in Baltimore at age 96. Often referred to as "the father of neuroscience," Mountcastle defied early skeptics by showing how cylinders of neurons, dedicated to specific tasks, work together. This month's Cerebrum *features remembrances from two colleagues influenced by Mountcastle—among the many who have gone on to make their own significant impacts in neuroscience.*

Laying the Foundation

By Mahlon R. DeLong, M.D.

Mahlon R. DeLong, M.D., is the Willliam P. Timmie Professor of Neurology at Emory University School of Medicine and former chair of the neurology department. Before moving to Emory in 1990, he was chair of neurology and professor of neurology and neuroscience at Johns Hopkins Hospital and School of Medicine, where he first began teaching in 1975. DeLong is an elected member of the Johns Hopkins Society of Scholars, the Institute of Medicine, and the National Association of the Academies of Science. He shared the 2014 Lasker-DeBakey Clinical Medical Research Award with Alim Benabid, M.D., Ph.D., and was awarded the 2013 Breakthrough Prize in the life sciences *"for defining the interlocking circuits in the brain that malfunction in Parkinson's disease—laying the groundwork for treatment of the disease by deep brain stimulation."* After earning his B.A. from Stanford University and attending graduate school in physiology there, he received his medical degree cum laude from Harvard Medical School.

I FIRST MET VERNON when I was a fellow at the National Institutes of Health (NIH) in Edward Evarts' laboratory. When I moved from the NIH to Johns Hopkins as a resident in neurology and then joined the faculty, Vernon generously provided me temporary research space in the Department of Physiology. The rich and stimulating environment and interaction with colleagues, fellows, and visiting faculty was unique. Vernon's early advice helped with my first NIH grant application, which, almost unthinkable in the current environment, was fully funded on the first round. Collaborative studies with Apostolos Georgopoulos, and discussions with colleagues including Tom Powell, were the highlights and the beginning of one the most productive periods of my career. It was most fortunate for me to have had Vernon's early and continued support.

Vernon, whose career at Johns Hopkins lasted more than 46 years, laid the foundations for modern neuroscience by his discovery of how brain cells in the cerebral cortex are organized. His eureka moment occurred in 1957, as he was charting the responses of individual brain cells in the sensory cortex of a cat to superficial tactile stimulation and pressure applied to the cat's paw. As he advanced a microelectrode from the surface of the brain to the deeper layers of the cortex, he observed that brain cells responding to a particular point on the skin, with responses to either superficial stimulation or deep pressure, were stacked on top of one another in narrow, vertical columns.

The finding of columnar organization stood in stark opposition to the prevailing view that the cortex, a clearly layered structure, was organized horizontally. Some even believed each cortical layer was responsible for different and unique functions. Neuroanatomists saw no anatomical evidence for vertical columns and considered it a radical hypothesis. So controversial was the finding that his co-investigators preferred not to be coauthors on the publication. But "Modality and Topographic Properties of Single Neurons of Cat's Somatic Sensory Cortex" is now unquestionably one of the most seminal and far-reaching publications in the field of neuroscience. His conclusion: "that the elementary pattern of organization in the cerebral cortex is a vertically oriented column or cylinder of cells."

Although many colleagues and physiologists as well as anatomists initially viewed this conclusion with skepticism, investigators studying other regions of the cortex involved with the processing of different modalities of sensory information confirmed Vernon's finding. He and his colleagues continued their careful investigations, first in the anesthetized cat and then the primate, finding even further evidence of separate channels for transmission of sensory information about stimulus type and location from the periphery to the parietal cortex and for columnar organization at the cortical level.

Although Vernon was most cited for his early work, it was his subsequent studies on the parietal cortex in the awake monkey that he considered most significant. The findings in recordings of single-cell activity in monkeys trained to perform movement tasks requiring attention to stimuli, decision making, and action in the form of movement were transformative. In his autobiography, prepared for the Society for Neuroscience in 1992, he commented that after failing to see any signs of the discriminatory process in the early stages of processing in the parietal cortex, they, "more or less in frustration," moved the recording to the posterior parietal cortex. What we saw that day determined my experimental life for 15 years. Neural responses to stimuli occurred only if the animal attended to them," he said.

Strikingly, Vernon and his colleagues observed a wide variety of motor, tactile, and visual responses in individual cells, reflecting the processes of sensory integration, attention, and decision making, observations—later substantiated through painstaking studies in trained primates. Again, they found that "neurons defined by these [different] functional properties are arranged in type-specific columns." Furthermore, they later showed that the functional properties of such neurons strikingly reflected the functions impaired following removal of the posterior parietal area in monkeys and also seen in patients with damage to this area.

Vernon's seminal contributions over the decades led the way for the fundamental view that brain functions are distributed, with multiple modules working in concert—receiving, processing, discriminating, and acting upon their inputs. The remarkable developments in modern systems of neuroscience rest upon the fundamental view of brain organization Vernon and his colleagues first elaborated.

Certainly, Vernon will be remembered most for his scientific contributions, but those who knew, worked, or trained with him personally attest to the depth and breadth of his knowledge, the clarity of his thinking and writing, his strong work ethic, and his fairness and passion for science and discovery. This passion is reflected poignantly in his autobiography as he reflects on the end of his investigative career in the early 1990s. It was then that he began examining the discrimination process linking the transitions from sensation to action in the cells of the motor cortex: "This was my last experience in laboratory research," he wrote. "I was nearly brokenhearted to leave it, for I found no greater thrill in life than to make an original discovery, no matter how small."

That passion may be a major factor in understanding his dedication and work ethic. He often finished his administrative work by 9 a.m., entered the lab, took a break for dinner at home with the family, and then returned to the lab and worked until midnight. He entered the darkened laboratory to the sound of action potentials displayed on the oscilloscope as an explorer in unexplored sacred space, the brain. As Robert LaMotte, a fellow with Mountcastle in the early studies of the parietal lobe, recalled "It was akin to going into a little submarine with him, like being the Jacques Cousteau of the cortex."

Vernon received nearly every major award in science, including the Albert Lasker Basic Medical Research Award. The Lasker award, often considered the "American Nobel," recognized him "for his original discoveries which illuminate the brain's ability to perceive and organize information, and to translate sensory impulses into behavior" and as "the intellectual progenitor of the many researchers at work in neuroscience today." David Hubel, who shared the Nobel Prize with Torsten Wiesel for their work on the processing of visual information, said in his acceptance speech: "[Vernon's] discovery of columns in the somatosensory cortex was surely the single most important contribution to the understanding of cerebral cortex since [Ramón y] Cajal." Sol Snyder, commenting on the prevailing dominance of genetics and molecular biology in a 2007 interview in *Hopkins Medicine* magazine, stated: "The more we know of individual genes that regulate brain function, the more it becomes clear that molecular biology is just the beginning—and we need to return to the lessons of Vernon Mountcastle to put it all together."

Snyder's words have gained traction as we enter the era of the BRAIN Initiative, with its emphasis on linking subcellular activity, single cells, and synapses to network function and, ultimately, to complex behavior. The burgeoning field of neuroscience mourns the loss of a pioneer who contributed so greatly to its development.

An Unforgettable Friend

By Guy McKhann, M.D.

Guy McKhann, M.D., the scientific advisor to the Dana Foundation, studies neurological outcomes following coronary artery bypass grafting, and the elucidation of the mechanism of a form of Guillen Barre Syndrome. He has also been active in defining the criteria for Alzheimer's disease. McKhann received his B.S. degree from Harvard University and obtained his doctoral degree from Yale Medical School. After working at the National Institute of Neurological Disorders and Stroke, he took his residency in pediatric neurology at Massachusetts General Hospital. His first academic position was at Stanford University, where he founded the pediatric neurology service. He then moved to Johns Hopkins University Medical Center, where he was the first director of the neurology department.

JOHNS HOPKINS AND THE WORLD OF NEUROSCIENCE lost one of their greats last month. At age 96, Vernon still thought and acted as he had his entire scientific life. Mahlon has done a wonderful job describing Vernon's scientific career. I will not add to that. I would, however, like to say more about Vernon as a person.

Vernon was born in Kentucky, but his family moved to Roanoke, Virginia, when he was three. Vernon was the ultimate Virginian: proud of his ancestors, proud of his country, and proud to represent them. He was extraordinarily bright, skipping two grades in primary school, going through Roanoke College in three years, and entering Johns Hopkins Medical School at 19. He was initially overwhelmed by the Hopkins environment, surrounded by classmates from much more prestigious academic backgrounds. Furthermore, his mother had wanted him to go to a Virginia medical school, rather than be surrounded by "all those Yankees." It wasn't long before Vernon was more than holding his own in his new environment—an environment he never left. He went on to spend his entire career at Hopkins, with the exception of a few years in the Navy during World War ll.

Vernon intended to be a neurosurgeon and had spent a year as a surgical intern before the Navy. On his return to Hopkins, he found that things had changed. Walter Dandy, a renowned neurosurgeon, had died, and no successor had been named. The head of surgery, Alfred Blalock, had deemed that no appointments in neurosurgery would be made until a new head of neurosurgery had been found. Vernon tried another tack, applying to a Hopkins spinoff program in neurosurgery at Duke University. The Duke program was also full. On an impulse, Vernon asked if spending a year in physiology at Hopkins with Philip Bard might qualify him for a residency. Barnes Woodall, head of neurosurgery at Duke, jumped at the chance, and said that if Bard would have Vernon around for a year, he could join the Duke program. Vernon returned to Baltimore for an interview with the unsuspecting Bard. It was one of the shortest on record:

BARD: "Do you think there is a psychological factor in motion sickness?"

VERNON: "No."

BARD: "Come in September."

Thus began the neurophysiology career of one of the leading neurophysiologists of all time, who was pleased to say he did his first experiment at age 28!

To digress for a moment, I have often wondered what would have happened if Vernon had gone into neurosurgery. I think he would have been superb. He had technical gifts, as was shown in his demanding physiological experiments. He took as long as was needed and focused on specific problems. He was a great admirer of Wilder Penfield at the Montreal Neurological Institute and could have been the American equivalent.

I knew Vernon well—he recruited me to Hopkins. Hopkins had never had a neurology department; at the time neurology was a small division of medicine. Vernon, along with Bob Cooke in pediatrics, Paul Talalay in pharmacology, and Dave Bodian in anatomy, lobbied for neurology becoming a department. Part of their argument was that in the previous ten years, not a single Hopkins medical student had gone into neurology. Additionally, neurology was more than an extension of medicine; it had a basic science requiring clinical application. The final selection process came down to two candidates: Richard Johnson and me. With the help of the superb dean at the time, Tommy Turner, two endowed professorships were obtained and Vernon recruited both of us. The development of the department to its current position as the number one department in the United States (according to *U.S. News and World Report*) is directly related to the direction that Dick and I provided, followed by our successors, Jack Griffin and Justin McArthur.

Vernon and I used to play tennis every Wednesday morning at 7 o' clock. Vernon was good—a human backboard in the sense that he hit everything back. If we played his game, good steady tennis, he would usually win. If I adapted to a game of drop shots and lobs, it was clear that Vernon didn't consider that tennis. I usually didn't, being somewhat ashamed of those tactics against an 80-year-old opponent. On one occasion a young man, whom I did not know, came up to me and asked: "Are you Dr. McKhann?" I said yes, and he went on to say that he wanted to thank me because he worked with Vernon in the lab and every time he won our tennis match, the lab was a much more peaceful place.

My other major interaction with Vernon was around the development of the Zanvyl Krieger Mind-Brain Institute. Vernon had first planted the idea with Steve Muller, president of Johns Hopkins at the time. Vernon proposed an institute that would focus on how the brain influences behavior. Muller bit and, after some start-up pains, the institute was formed. Vernon talked me into giving up the direction of the neurology department to Dick Johnson and becoming the institute's first head. A key component of launching the institute was moving Vernon's former colleagues there; the purpose was to make it a center of systems for neuroscience. The institute has done well despite the death of two of its stars, Ken Johnson, who succeeded me as director, and Steve Hsiao, who was its scientific director. The present director, Ed Connor, is continuing Vernon's considerable legacy.

I cannot finish without remarking about Vernon's remarkable wife, Nancy. She is 92 and still going strong. Nancy balanced Vernon's life, seeing to the kids, taking in foreign graduates and visitors, teaching school, and also playing tennis into her late 80s—all this with Virginian charm. Nancy is a steel magnolia if there ever was one, and I mean that in a most positive way.

This has been a difficult piece for me to write so soon after Vernon's death. I say goodbye to a first-rate scientist, human being, and friend.

36 • Cerebrum 2015

4

The Darkness Within

Individual Differences in Stress

By George F. Koob, Ph.D.

George F. Koob, Ph.D., is director of the U.S. National Institute on Alcohol Abuse and Alcoholism at the National Institutes of Health. (He is currently on a leave of absence from the Committee on the Neurobiology of Addictive Disorders at The Scripps Research Institute and Departments of Psychology and Psychiatry and Skaggs School of Pharmacy and Pharmaceutical Sciences at the UC, San Diego.) Koob, an authority on drug addiction and stress, has contributed to the understanding of the neurocircuitry associated with the acute reinforcing effects of drugs of abuse and the neuroadaptations of the reward and stress circuits associated with the transition to dependence. He is co-author of *Drugs, Addiction and the Brain*, and earned his Ph.D. in behavioral physiology at Johns Hopkins University.

Editor's Note: Numerous factors make us react to situations differently: age, gender, education, relationships, socioeconomic status, environment, cultural background, life experience. But as our author describes, biological bases, such as the way genetics and neurochemicals affect our brains, are providing insight into addiction, posttraumatic stress disorder, and other stresses that he calls "an intimate part of modern life."

STRESS IS EVERYWHERE. It is an intimate part of modern life. But what is stress? How does the brain process the feeling as a "stress system"? What chemicals in our brains mediate the stress response, and, most important, can we control it? Moreover, what conveys individual differences in stress responsivity that leave some of us vulnerable to stress disorders and others resilient? When does stress go rogue and produce psychopathology? And why do I think of it as the "dark side" of reward pathways in the brain?

My hypotheses are that individual differences in stress vulnerability and resilience are key determinants of the development of posttraumatic stress disorder (PTSD) and addiction, and these differences derive from the neurocircuitry of our emotional dark side. I'll take you through this neurocircuitry to explain what I mean.

What Is Stress?

Stress can be classically defined as "the nonspecific (common) result of any demand upon the body"[1] or, from a more psychological perspective, "anything which causes an alteration of psychological homeostatic processes."[2] Historically, the physiological response that is most associated with a state of stress is an elevation of chemicals called glucocorticoids that help control inflammation. Glucocorticoids are derived from the adrenal cortex, a gland situated above the kidneys, and glucocorticoid elevations were thought to be controlled by the brain's hypothalamus, a region that is associated with emotion. Maintaining psychological homeostasis, therefore, involves responses among the nervous, endocrine, and immune systems. This nexus is referred to as the hypothalamic-pituitary-adrenal (HAP) axis.

Efforts to identify processes involved in disrupting psychological homeostasis began while I was a staff scientist at the Arthur Vining Davis Center for Behavioral Neurobiology at the Salk Institute in California. My colleagues Wylie Vale, Catherine Rivier, Jean Rivier, and Joachim Spiess first demonstrated that a peptide called corticotropin-releasing factor (CRF) initiates the HPA axis's neuroendocrine stress response. Research showed that CRF emanated from a part of the hypothalamus called the paraventricular nucleus, which is the primary controller of the hypothalamic-pituitary-adrenal axis. When the hypothalamus releases CRF, it travels through blood vessels to the pituitary gland, located at the base of the brain. There, CRF binds to receptors located in the anterior part of this gland to release adrenocorticotropic hormone (ACTH) into the blood stream.[3]

ACTH in turn travels to the cortex of the adrenal gland to release glucocorticoids. Glucocorticoids, in turn, synthesize glucose to increase energy used by the brain, and glucocorticoids also decrease immune function by blocking "proinflammatory" proteins that ordinarily produce inflammation. Together these responses facilitate the body's mobilization in response to acute stressors. Indeed, acute and chronic glucocorticoid responses differentially affect brain function, with acute high-dose glucocortoids imparting a protective effect.[4]

Fight or Flight?

When faced with stressors, what determines whether we fight or flee? The human brain's "extended amygdala" processes fear, threats, and anxiety (which cause fight or flight responses in animals)[5,6] and encodes negative emotional states. Located in the lower area of the brain called the basal forebrain, the extended amygdala is composed of several parts, including the amygdala and nucleus accumbens.[7] This system receives signals from parts of the brain that are involved in emotion, including the hypothalamus and, most important for this examination, the prefrontal cortex. Extended amygdala neurons send axons or connections heavily to the hypothalamus and other midbrain structures that are involved in the expression of emotional responses.[7,8]

In psychopathology, dysregulation of the extended amygdala has been considered important in disorders related to stress and negative emotional states. These disorders include PTSD, general anxiety disorder, phobias, affective disorders, and addiction.[9,10] For example, animals exposed to a stressor will show an enhanced freezing response to a conditioned fear stimulus, an enhanced startle response to a startle stimulus, and avoidance of open areas, all of which are typical responses to an aversive stimulus and are mediated in part by the extended amygdala.

The Neurochemical Mediators

Why then do individual responses to stress differ? Two important neurochemical systems are involved and help answer this question. The first one is CRF, the neurochemical system mentioned above. It turned out CRF is also a major component of the extended amygdala and works to effect behavioral changes.

While the glucocorticoid response mobilizes the body for physiological responses to stressors, CRF mobilizes the body's behavioral response to stressors via brain circuits outside the hypothalamus. One of my first eureka moments was when my laboratory helped demonstrate initially that CRF mediates not only physiological and hormonal responses to stressors but also behavioral responses.

In our first study, I injected the newly discovered CRF peptide into the brain in rats and observed very peculiar behavioral hyperactivity. The rats climbed all over the wire-mesh testing cages, including the walls. I called Wylie Vale over to observe the animals because they seemed to be levitating. We subsequently showed that injecting CRF into the rats' brains produced a pronounced hyperarousal in a familiar environment but a pronounced freezing-like response in a novel stressful environment.[11] Subsequent work showed that the extended amygdala mediates such responses to CRF and fear and anxiety in general. When agents were used to block CRF receptors from binding CRF, antistress effects occurred, confirming that the release of naturally produced CRF is central in behavioral responses to stressors.[12] Equally intriguing, in chronic prolonged stress, glucocorticoids stimulate CRF production in the amygdala while inhibiting it in the hypothalamus, suggesting a means of protecting the body from high chronic exposure to glucocorticoids by shutting off the HPA axis but driving the extrahypothalamic CRF stress system.

The other key neurotransmitter system involved in individual differences in stress responsiveness is called the dynorphin-kappa opioid system (also located in the extended amygdala). This system is implicated in effecting negative emotional states by producing aversive dysphoric-like effects in animals and humans.[13] Dysphoria is a negative mood state, the opposite of euphoria. Dynorphins are widely distributed in the central nervous system.[14] They have a role in regulating a host of functions, including neuroendocrine and motor activity, pain, temperature, cardiovascular function, respiration, feeding behavior, and stress responsivity.[15]

In addition to these two neurochemical systems, we now know that other neurochemical systems interact with the extended amygdala to mediate behavioral responses to stressors. They include norepinephrine, vasopressin, hypocretin (orexin), substance P, and proinflammatory cytokines. Conversely, some neurochemical systems act in opposition to the brain stress systems. Among these are neuropeptide Y, nociceptin, and endocannabinoids. A combination of these chemical systems sets the tone for the modulation of emotional expression, particularly negative emotional states, via the extended amygdala.[16]

Psychopathology and Stress Systems

How are stress systems involved in PTSD? PTSD is characterized by extreme hyperarousal and hyperstress responsiveness. These states contribute greatly to the classic PTSD symptom clusters of re-experiencing, avoidance, and arousal. Perhaps more insidious, about 40 percent of people who experience PTSD ultimately develop drug and alcohol use disorders. Data suggest that the prevalence of an alcohol use disorder in people with PTSD may be as high as 30 percent.[17] The major model of PTSD neurocircuitry evolved from early animal work on fear circuits,[18] which suggested that brain stress systems are profoundly activated in the extended amygdala.

PTSD patients exhibit abnormally high glucocorticoid receptor sensitivity. This hypersensitivity results in excessive suppression of the HPA axis through corticosteroid negative feedback.[19] Research has found that military participants who developed high levels of PTSD symptoms after deployment tended to be those who had significantly higher glucocorticoid receptor expression levels before deployment.[20] Another key preclinical study showed that strong activation of CRF receptor signaling in animal models can induce severe anxiety-like and startle hyperreactivity that corresponds to the severe anxiety and startle reactivity seen in patients with PTSD.[21] Research also has demonstrated that patients with severe PTSD exhibit overly active brain CRF neurotransmission, measured by increases in CRF in their cerebrospinal fluid.[22]

While data on PTSD and the dynorphin-kappa system are limited, significant data suggest that brain kappa-opioid receptors play an important role in mediating stress-like responses and encoding the aversive effects of stress.[13] An exciting recent imaging study with a kappa-opioid tracer showed decreased kappa-opioid binding in the brain in PTSD patients. This finding suggests increased dynorphin release in patients who are clinically diagnosed with PTSD.[23]

From a neurocircuitry perspective, functional imaging studies of patients with PTSD show that the amygdala is hyperactive, while the ventromedial prefrontal cortex (PFC) and inferior frontal gyrus area show reduced activity.[24] These findings suggest that the ventromedial PFC no longer inhibits the amygdala. This loss of inhibition in turn drives increased responses to fear, greater attention to threatening stimuli, delayed or decreased extinction of traumatic memories, and emotional dysregulation.[25]

One attractive hypothesis for the functional neurocircuitry changes that occur in PTSD suggests a brain-state shift from mild stress (in which the PFC inhibits the amygdala) to extreme stress (in which the PFC goes offline and the amygdala dominates; see Figure 1).[26] Under this paradigm (rubric means "a standard of performance for a defined population"), relative dominance by the cerebral cortex conveys resilience, and relative dom-

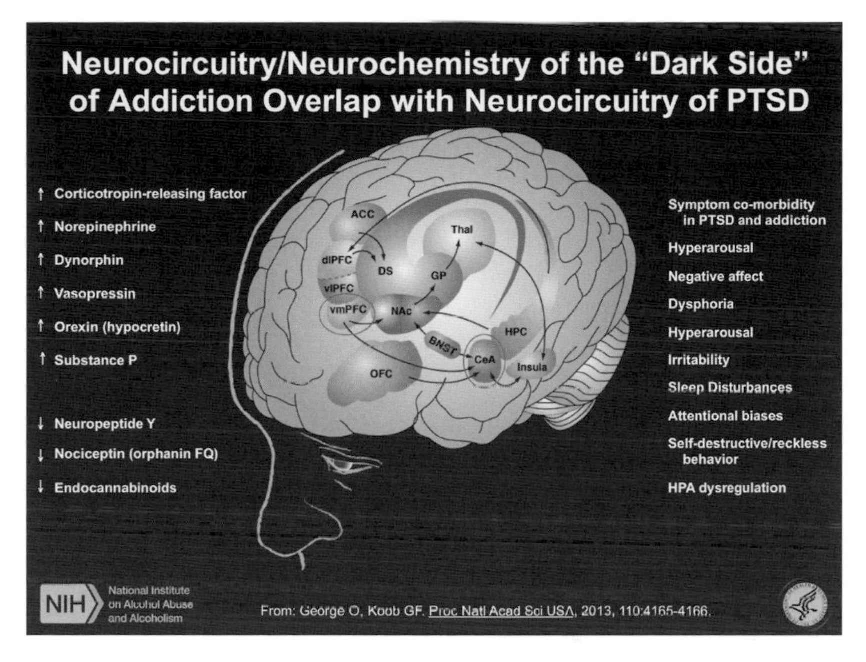

Figure 1. Common neurocircuitry in addiction and posttraumatic stress disorder (PTSD) with a focus of prefrontal cortex (PFC) control over the extended amygdala. The medial PFC inhibits activity in the extended amygdala, where key stress neurotransmitters mediate behavioral responses to stressors and negative emotional states. Key neurotransmitters include corticotropin-releasing factor (CRF) and dynorphin but also other stress and antistress modulators. Notice a significant overlap in the symptoms of PTSD and the withdrawal/negative affect stage of the addiction cycle.

inance by the amygdala conveys vulnerability.[26] Delving further into the effects of prefrontal control, two related studies showed that ventromedial PFC activation correlates with the extinction of fear, whereas amygdala activation by the dorsal anterior cingulate cortex (ACC) correlates with a failure to eliminate fear.[27,28]

The Paradoxical 'Darkness Within'

I often tell people that I spent the first 15 years of my career studying why we feel good and the most recent 15 years studying why we feel bad. However, these two emotional states are intimately linked, which raises the seemingly contradictory possibility that excessive activation of the reward system can lead to stress-like states that, in their severest form, resemble PTSD. So how did I get to the "dark side"? Well, by first studying the "light side," or how drugs produce their rewarding effects.

My research team and others hypothesized that addiction involves three stages that incorporate separate but overlapping neurocircuits and relevant neurotransmitter systems: binge/intoxication, withdrawal/negative affect, and preoccupation/anticipation or "craving."[29,30] The binge/intoxication stage involves the facilitation of incentive salience (the linking of previously neutral stimuli in the environment to rewards to give those stimuli incentive properties), mediated largely by neurocircuitry in the basal ganglia. The focus is on activation of the "reward" neurotransmitters dopamine and opioid peptides that bind to mu-opioid receptors in the brain. Early work in the addiction field showed that the nucleus accumbens was a key part of this neurocircuitry that mediates the rewarding properties of abused drugs.

Franco Vaccarino and I showed that we could block heroin self-administration when we injected minute amounts of methylnaloxonium, which blocks opioid receptors, into animals' nucleus accumbens.[31] Subsequently, several classic human imaging studies showed that intoxicating doses of alcohol result in the release of dopamine and opioid peptides in the nucleus accumbens.[32,33] We now know that activation of the nucleus accumbens leads to the recruitment of basal ganglia circuits that engage the formation and strengthening of habits. This process is hypothesized to reflect the beginning of compulsive-like responding for drugs—in other words, addiction.

An experiment that turned out exactly the opposite of what I had predicted is the second reason I landed on addiction's dark side. Tamara Wall, Floyd Bloom, and I set out to identify which regions of the brain mediate physical withdrawal from opiates. We began by training opiate-dependent rats to work for food. Then we disrupted their food-seeking behavior by injecting them with naloxone. This drug precipitated withdrawal, producing a malaise- and dysphoric-like state; as a result, the rats stopped pressing the lever. Thus far, we had successfully replicated original findings.[34] We then set out to inject methylnaloxonium, a drug that blocks opioid receptors in brain areas previously implicated in physical withdrawal from opiates. We injected this drug because it was a naloxone analog that would spread less in the brain and precipitate "local" withdrawal as measured by a decrease in lever pressing for food.

We speculated that the most sensitive brain areas to produce a decrease in lever pressing would be the periaqueductal gray and medial thalamus because they had been shown to mediate physical withdrawal from opiates. However, injections into the periaqueductal gray and medial thalamus were ineffective in decreasing lever pressing for food. Instead, injections into the nucleus accumbens proved effective—so effective that we had to drop the dose. Even at a very low dose, we saw some modest effect in decreasing lever pressing for food.[35] It then dawned on me that the same brain region responsible for making you feel good also made you feel bad when you became dependent (addicted). This epiphany led me to devote the rest of my career to trying to understand exactly how such opposite reactions that occur during withdrawal, termed opponent processes, are mediated.

This observation led me to a completely new conceptualization of the withdrawal/negative affect stage of addiction. I concluded that this stage is characterized not only by drug-induced specific "physical withdrawal" but also common drug-induced "motivational" withdrawal, characterized by dysphoria, malaise, irritability, sleep disturbances, and hypersensitivity to pain. (These symptoms are virtually identical to the hyperarousal/stress symptoms seen in PTSD; see Figure 1).

Two processes were subsequently hypothesized to form the neurobiological basis for the withdrawal/negative affect stage. One is the loss of function in the reward systems in the medial part of the nucleus accumbens of the extended amygdala. This reward system loss is mediated by a loss of function in dopamine systems. The other process is the recruitment of brain stress systems in other parts of the extended amygdala (notably, the central nucleus of the amygdala), including recruitment of the neurochemical systems CRF and dynorphin.[36,37] The combination of decreases in reward neurotransmitter function and recruitment of brain stress systems provides a powerful motivation for reengaging in drug taking and drug seeking.

Yet another breakthrough came when my laboratory first realized the dramatic role of CRF in compulsive alcohol seeking, via the amelioration of anxiety-like responses when a CRF receptor antagonist or receptor blocker was used to block the anxiety-like responses of alcohol withdrawal.[38] Subsequently, we showed that acute alcohol withdrawal activates CRF systems in the central nucleus of the amygdala.[39] Moreover, in animals we found that site-specific injections of CRF receptor antagonists into the central nucleus of the amygdala or systemic injections of small-molecule CRF antagonists reduced the animals' anxiety-like behavior and excessive self-administration of addictive substances during acute withdrawal.[12,40] Perhaps equally compelling, Leandro Vendruscolo and I recently showed that a glucocortoid receptor antagonist could also block the excessive drinking during acute alcohol withdrawal, linking sensitization of the CRF system in the amygdala to chronic activation of the HPA glucocorticoid response.[41]

But how is excessive activation of the reward system linked to activation of the brain stress systems? Seminal work by Bill Carlezon and Eric Nestler showed that the activation of dopamine receptors that are plentiful in the shell of the nucleus accumbens stimulates a cascade of events that ultimately lead to changes in the rate of DNA transcription initiation and alterations in gene expression. Ultimately, the most notable alteration is activation of dynorphin systems. This dynorphin system activation then feeds back to decrease dopamine release.[37] Recent evidence from my laboratory and that of Brendan Walker suggests that the dynorphin-kappa opioid system also mediates compulsive-like drug responses (to methamphetamine, heroin, nicotine, and alcohol); this response is observed in rat models during the transition to addiction. Here, a small-molecule kappa-opioid receptor antagonist selectively blocked the animals' development of compulsive drug self-administration.[42-45] Given that the activation of kappa receptors produces profound dysphoric effects, this plasticity within the extended amygdala may also contribute to the dysphoric syndrome associated with drug withdrawal that is thought to drive the compulsive responses mediated by negative reinforcement.[46]

Yet another pleasant surprise was the realization that the preoccupation/anticipation, or "craving," stage in alcoholism mediates the dysregulation of executive control via prefrontal cortex circuits. Importantly, these circuits can become a focal point for individual differences in vulnerability and resilience. Many researchers have conceptualized two generally opposing systems, a "Go" system and a "Stop" system, where the Go system engages habitual and emotional responses and the Stop system brakes habitual and emotional responses. The Go system circuit consists of the anterior cingulate cortex and dorsolateral PFC, and it engages habit formation via the basal ganglia. The Stop system circuit consists of the ventromedial PFC and ventral anterior cingulate cortex and inhibits basal ganglia habit formation, as well as the extended amygdala stress system. People with drug or alcohol addiction experience disruptions of decision making, impairments in the maintenance of spatial information, impairments in behavioral inhibition, and enhanced stress responsivity, all of which can drive craving. More important, this Stop system controls the "dark side" of addiction and the stress reactivity observed in PTSD.

This realization was brought home to me when my colleague Olivier George and I showed that, even in rats that simply engaged in the equivalent of binge drinking, there was a disconnection of the frontal cortex's control over the amygdala but not nucleus accumbens.[47] These results suggest that early in excessive alcohol consumption, a disconnect occurs in the pathway between the PFC and the central nucleus of the amygdala, and this disconnect may be key to impaired executive control over emotional behavior.

Evidence for a Genetic/Epigenetic Mechanism

I suspect that the neurocircuitry focus on the frontal cortex and amygdala in the development of PTSD and addiction will reveal targets for individual differences in vulnerability and resilience. Human imaging studies have established that reduced functioning of the ventromedial PFC and anterior cingulate cortex and increased functioning of the amygdala are reliable findings in PTSD.[26] Similarly, drug addiction also has been associated with general reduced function of the ventromedial PFC.[48] So what is the contribution of the ventralmedial PFC and anterior cingulate cortex in stress and negative emotional states associated with craving, particularly given what we already know in PTSD? Considering the high co-occurrence of substance abuse and PTSD and the key role of the PFC in controlling the stress systems, the dysregulation of specific subregions of the PFC may be involved in both disorders.

Converging evidence in humans suggests major individual differences in the response of the extended amygdala to emotional stimuli, particularly those considered stressful, and in vulnerability to PTSD and addiction. Research has demonstrated that the central nucleus of the amygdala (the dorsal amygdala in humans) is involved in the conscious processing of fearful faces in healthy volunteers and, more important, that individual differences in trait anxiety predicted the response of a key input to the central nucleus of the amygdala, the basolateral amygdala, to unconsciously processed fearful faces.[49] Moreover, a landmark study that used positron emission tomography showed that the amygdala is activated in cocaine-addicted individuals during drug craving but not during exposure to non-drug-related cues.[50]

Similarly, changes in frontal cortex function can convey individual differences in vulnerability and resilience. In one prospective study that was conducted following the 9.0 Tohoku earthquake in Japan in 2011, participants who had higher gray matter volume in the right ventral anterior cingulate cortex were less likely to have developed PTSD-like symptoms.[51] The degree of improvement in symptoms after cognitive behavior therapy was positively correlated with increases in anterior cingulate cortex activation.[52] In contrast, other studies have shown that people with PTSD and their high-risk twins show greater resting brain metabolic activity in the dorsal anterior cingulate cortex compared with trauma-exposed individuals without PTSD, suggesting that increased dorsal anterior cingulate cortex activity may be a risk factor for developing PTSD.[53]

But what molecular neurobiological changes drive these circuit changes? Genetic studies have shown that 30 to 72 percent of the vulnerability to PTSD and 55 percent of the vulnerability to alcoholism can be attributed to heritability. Most would argue that the genetic influences of both disorders stem from multiple genes, and the candidate-gene approach has not yet identified major genetic variants that convey vulnerability to PTSD. However, in two scholarly reviews, at least 17 gene variants were associated with PTSD and many others with alcoholism.[26] Overlapping genes that have been identified in both disorders include gamma-aminobutyric acid, dopamine, norepinephrine, serotonin, CRF, neuropeptide Y, and neurotrophic factors, all of which are relevant to the present hypothesis.

From an epigenetic perspective, some genes may be expressed only under conditions of trauma or stress, and these environmental challenges can modify genetic expression via DNA methylation or acetylation. Both PTSD and alcoholism show epigenetic changes that suggest an increased regulation in genes related to the stress system.[54,55] For PTSD, one gene that has been implicated in epigenetic modulation is SLC6A4, which regulates synaptic serotonin reuptake and appears to have a central role in protecting individuals who experience traumatic events from developing PTSD via high methylation activity.[56] For alcoholism, histone deacetylase (HDAC) has been implicated in an epigenetic modulation. This gene is involved in the activity-dependent regulation of brain-derived neurotrophic factor (BDNF) expression in neurons. Alcohol-preferring rats with innate higher anxiety-like responses showed higher HDAC activity in the central nucleus of the amygdala. Knockdown of a specific HDAC called HDAC2 in the central nucleus of the amygdala increased BDNF activity and reduced anxiety-like behavior and voluntary alcohol consumption in a selected line of rats that were bred for high alcohol preference.[57]

Thus, altogether, my hypothesis is that individual differences in stress vulnerability and resilience, which are key determinants of the development of PTSD and addiction, derive from the neurocircuitry of our emotional "dark side." The origins of activation of the dark side involve both hyperactivity of the extended amygdala (dynorphin and CRF driven by excessive drug use) and reduced activity of the medial PFC (driven by excessive drug use and brain trauma). New advances in our understanding of the neurocircuitry of the dark side and identification of epigenetic factors that weight the function of these circuits will be the key to precision medicine for the diagnosis and treatment of these disorders.

5

Tau-er of Power

By Kenneth S. Kosik, M.D.

Kenneth S. Kosik, M.D., is the Harriman Professor of Neuroscience Research and co-director of the Neuroscience Research Institute at the UC, Santa Barbara. Kosik's work with early-onset familial Alzheimer's disease at Columbia University was the basis for a novel prevention trial to treat Alzheimer's disease. His was one of several groups that discovered tau protein in the Alzheimer's neurofibrillary tangle and followed up with many studies on the biology and pathobiology of tau. Kosik received a B.A. and M.A. in English literature from Case Western Reserve University in 1972 and an M.D. from the Medical College of Pennsylvania in 1976. He served as a resident in neurology at Tufts-New England Medical Center and was chief resident in 1980. Since 1980, he has held a series of academic appointments at the Harvard Medical School and achieved the rank of full professor there in 1996. He also held appointments at McLean Hospital, Brigham and Women's Hospital, Massachusetts General Hospital, and the Dana-Farber Cancer Institute. In 2004, he accepted the appointment at the UC, Santa Barbara.

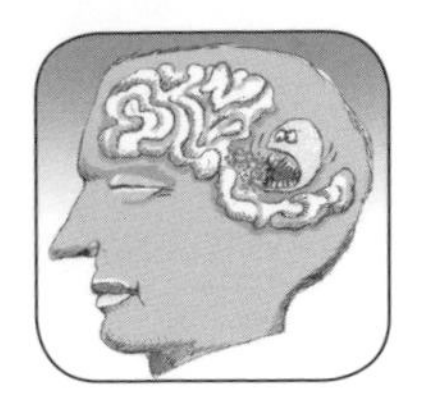

Editor's Note: Tau protein helps nerve cells in the brain maintain their function and structure. When tau turns toxic, replicates, and spreads, neurons misfire and die. If neuroscientists can pinpoint the reasons for toxicity, identify what our author calls "a staggering number of possible modified tau states," and find a way to block tau's movement from cell to cell, then progress can be made in fighting any number of neurological disorders linked to this protein, including frontotemporal dementia, chronic traumatic encephalopathy (CTE), and Alzheimer's disease.

MENTION ANY DISEASE and a few questions immediately come to mind. Chief among them: Who is vulnerable, and how does it occur? If it's an infectious disease, it may spread through the air or by touch. But the presiding dogma for most of modern biomedical history tells us that the transmissible agents contain nucleic acid and that replication is inextricably bound to DNA or RNA. As information vehicles, these molecules power the dual aspects of evolution: constancy and change. The constancy of faithful replication allows the inheritance of traits that allow life forms (including classical disease culprits) to survive. Through change, genes ensure adaptability to a complex environment. These properties confer pathogenicity by enabling them to prey upon infectious agents and adapt to their hosts.

But a parallel universe of disease transmission ignores these rules. This is the world of prions. In this world, the agent of transmission is a protein and the information lies in the vast shape of the space within which proteins fold. How big is that space? To get a handle on the size of chemical space, the chemistry blogger Derek Lowe quoted Douglas Adams in *The Hitchhiker's Guide to the Galaxy*: "Space is big. Really big. You just won't believe how vastly, hugely, mind-boggling big it is." The immense variety within stretches of ATGCs (the chemicals adenine, cytosine, guanine, and thymine) that build the language of the genetic code pales beside the alphabet of the periodic table with its spelling and grammatical rules of chemical bond formation and compound stability.

Tau, a Protean Protein

Tau is a normal protein used in neurons to shape a dynamic system of tracks that traffic goods to various destinations inside the cell. Tau is one small fraction of the bewilderingly large space within which all proteins fold. But like a fractal image, zooming in on tau opens up a chemical space full of molecular crevices and passageways that mirror the far greater universe from which tau is but one tiny part. Like the early sea god, Proteus, tau can take many forms. It was said that Proteus had the ability to foretell the future but would change his shape to avoid doing so. Indeed, some forms of tau are harbingers of a future with neurodegeneration (more on this point later). The chemical space that tau occupies in the nervous system must first be catalogued according to the six different molecular isoforms of the tau protein, which many different chemical processes can modify. These six isoforms each have slightly different sequences with small stretches of amino acids either left in or out.

Even a cursory quantitative knowledge of the staggering number of possible modified tau states does not exist. Beyond these molecular states lies a far more extensive terrain of folding patterns, termed conformations (think of multiple ways to a crumple a piece of paper). And the only constraint on these conformations is the time they dwell in any one of them. Some shapes are stable only for a few milliseconds, and how readily the protein can assume a particular shape among all possible shapes is called the kinetic landscape.

Within the enormous realm of protein shapes, those that pertain to tau are little studied and poorly understood. Tau is in a unique class of proteins called intrinsically disordered proteins (IDPs). In contrast to enzymes, for example, which adopt a precise three-dimensional structure to facilitate catalysis, IDPs lack a unique three-dimensional structure and do not exhibit any stable secondary structure in the free form. They can adopt a wide variety of extended and compact conformations that facilitate many vital physiological functions by folding after they bind to targets. As with other IDPs, enzymes act on tau proteins twice as often as on other proteins and can alter their binding properties. Among these enzymes are a category called kinases that add a phosphate to a protein. IDPs are, on average, substrates of twice as many kinases as structured proteins.[1] Tau is particularly singled out as a substrate of multiple kinases, and some investigators believe that the multiple phosphates that decorate tau contribute to its tendency toward misfolding.

IDPs use their lack of structure to their advantage. The protean shapes that these proteins can assume provide a larger interaction surface area than globular proteins of a similar length. The variety of shapes exposes short linear peptide motifs that serve as molecular recognition features, and thereby allow IDPs to scaffold and interact with numerous other proteins. It enables diverse post-translational modifications that facilitate regulation of their function and stability in a cell; and by folding upon binding, IDPs can interact with their targets with relatively high specificity and low affinity. These features are ideal for recognizing partners to interact with and for coordinating regulatory events in space and time.[2] However, these properties require that cells assiduously monitor these proteins because they are potentially dangerous and capable of inflicting damage to cells by binding to each other.

Indeed, cells tightly regulate IDPs throughout, from transcript synthesis to protein degradation. Among the means that cells can use to adjust the levels of proteins is by the process of post-transcriptional regulation through microRNAs, which have been credited with helping to maintain tau homeostasis.[3] MicroRNAs target specific messenger RNAs, which encode proteins and fine-tune the amount of protein that gets translated from the messenger RNA.

Tau Aggregation

Paradoxically, simple overexpression of tau in a variety of cell types, including neurons in laboratory tissue culture, does not result in the replication and spread of tau (aggregation), even with high expression levels. More often tau overexpression induces the rampant assembly of a structure called microtubules, which are the cell's railroad tracks that ship cargo to different locations in the cell. Normally, tau promotes the assembly of microtubules, with the goal of building a protrusion from the cell that will become an axon. These long cylindrical structures use their microtubule railroads to carry cargo over the vast distances that axons traverse, such as the axons that travel from the lower spinal cord all the way to the muscles of the big toe. In the brain, axons connect the two hemispheres and are key elements of brain circuitry. When tau is overexpressed in nonneural cells, the out-of-control microtubule assembly results in numerous microtubule bundles spiraling around the perimeter of the cell, which is unable to form a protrusion in this foreign environment. Curiously, when tau is expressed in a type of cell called Sf9, taken from the ovary of an insect, it acquires many of the modifications seen in neurofibrillary tangles. With tau, these cells extrude a single, very long process that resembles an axon in its shape but has none of the electrical conduction properties of an axon.[4] Thus, tau makes an axon ghost in these cells. One conclusion from these studies is that simply lowering tau levels across the board is not the most strategic way to approach therapeutics for diseases termed tauopathies. Nor is simply overexpressing tau the way to make aggregates.

Some proteins can misfold into shapes called prions (as noted above) that have the unusual property of inducing other copies of the same protein to misfold similarly. The prion guides similar conformations in additional copies of the same protein. The prototypical prion, known as PrP, is responsible for several human diseases: Creutzfeldt-Jakob disease, Gerstmann–Sträussler–Scheinker syndrome, fatal familial insomnia, kuru, and bovine spongiform encephalopathy (also known as mad cow disease). Each of these diseases has very different clinical presentations, and they are distinguished from other infectious diseases mainly by their mode of transmission. Given that PrP causes all of these conditions and others found in nonhuman species, the idea that prion strains with distinct phenotypes exist has gained traction and experimental validation. This view suggests that a subset of PrP shapes is transmissible. Depending on the particular folding of a strain, a specific phenotype or species predisposition arises. Faithfully propagating strains, therefore, is a prerequisite for clinically defined presentations

What has been peculiar since the discovery of PrP is that only one human prion is known. Certainly other proteins must have kinetic troughs into which a protein can fall, get stuck, and spread its conformation to others. But only in yeast was a similar phenomenon clearly observed, and remarkably the prion state of the yeast protein confers a selective advantage.[5] For many years, investigators nibbled at the concept of prions to explain numerous neurodegenerative diseases in which a misfolded protein aggregates and remains trapped inside the cell. Among these conditions are the tau aggregates in the tauopathies, synuclein aggregates in Parkinson's disease, huntingtin aggregates in Huntington's disease, TDP-43 in frontotemporal lobar degeneration and amyotrophic lateral sclerosis, and transthyretin in familial amyloidotic neuropathy. Prions conceptually unify the neurodegenerative diseases, which otherwise lack a fundamental disease mechanism akin to a virus or a malignant cell or an autoimmune process in other disease categories. At the core of neurodegeneration lies an unknown process as mysterious as aging.

Protein aggregates are puzzling entities that lie on the wrong side of what in other contexts we call protein-protein interactions. Normal cellular function requires that proteins bind to each other as pairs or somewhat larger complexes. On the other hand, when tau aggregates, it grows massively, with molecular weights in the millions, and may capture other proteins within the aggregate.[6] In many cases, including tau inclusions, some feature of tau that is predisposed to self-assemble triggers recognition by a class of enzymes called ubiquitin ligases, which mark the protein for degradation. These enzymes enable a set of reactions that attach a chain of ubiquitin peptides to a protein. The ubiquitins tell the cell that the protein to which they are attached is trash and should be thrown away in the cell's trash can (called the proteasome). However, for unknown reasons, when ubiquitins attach to tau, it does not get degraded in the proteasome. Instead, it gets stuck in the cell as an aggregate. Maybe the size of the aggregate does not fit into the opening of proteasome, which is shaped like a barrel into which proteins enter for destruction inside.

The Spreading of Tau

In addition to the ability to evade the cellular surveillance systems that rid the cell of damaged proteins, prions have the even more insidious property of spreading to contiguous cells. Postmortem brain extracts from humans who had died with various tauopathies were injected into the hippocampus and cerebral cortex of mice and could be propagated between mouse brains.[7] The passage of a particular tau conformer appeared to faithfully replicate the pathology specific to each one of three clinically distinct types of tauopathies.

Faithful replication of tau aggregates was also demonstrated at the cellular level. Strains differed with respect to the morphology, size, and subcellular localization of the aggregates as well as their sedimentation profile, seeding capacity, protease digestion patterns, and toxicity.[8] Whether these features provide reliable "bar codes" for diagnosis will be known in the next few years.

Mice engineered to express a pathogenic human tau transgene in the entorhinal cortex, a highly vulnerable region involved in the sense of smell where tau pathology often begins, can spread to neuroanatomically connected regions.[9,10] Furthermore, mouse tau was bound up in the aggregates, suggesting that the pathological human tau induced normal mouse tau to misfold. Among the cells into which tau spread were dentate granule cells that are separated from the entorhinal cortical cells by a synapse. Whether spread occurs transsynaptically remains a fascinating, open question.

These findings and related tissue culture system studies point to four areas for deeper investigation: (a) how tau exits through the cell's membrane; (b) whether the existence of tau in an extracellular compartment offers a rationale for removing transmissible tau with an antibody; (c) how tau passes into the membrane of into a neighboring cell; and (d) what potential way stations—such as microglia—could interface with tau during its spread.[11]

Tau Seeding

So let's return to the myriad shapes tau can assume. Among these shapes are a few that expose some sticky surfaces normally kept folded and concealed within the protein. When this happens, tau can bind to another tau protein in a process called "seeding." As more and more tau proteins join the pack, eventually the aggregate becomes quite large and often appears like a fibril. Proteins, such as tau, that can wiggle and squirm in numerous ways are IDPs, and they have generated great interest in the scientific community. By changing their shape rapidly, they present different surfaces to other proteins and engage in a variety of binding interactions for their normal function.

Normally tau transitions on and off microtubules by folding in different ways and, in so doing, can stabilize and elongate the microtubule. However, during these on-off transitions, there are vulnerable moments with the potential for misfolding. Or, while tau is being synthesized from its mRNA, inopportune moments might allow tau to fold in a way that permits binding to another tau and seed the growth of a tau aggregate.

An aggregation-prone motif observed in tau is called a steric zipper, in which a pair of sheetlike sequences is held together by the interlinking of small projections.[12] Researchers have designed inhibitors that slow tau fibrillation by targeting a set of six amino acids in tau involved in this interaction.[13] Many unlikely events have to occur together for tau to form an aggregate: It must drop into a rare conformation or shape, it must remain in that shape for sufficient time to seed assembly of other tau proteins, and other tau proteins must be sufficiently close to serve as substrates upon which misfolded tau can template its nefarious shape.

Any mistakes in folding are monitored by classes of proteins called chaperones and co-chaperones. These proteins can refold a protein correctly or, if irreversibly damaged, direct the protein to the proteasome for complete degradation. The tethering of one such complex called BAG2/Hsp70 to the microtubule may provide a protective capture function for misfolded tau.[14]

Misfolded conformations may occur rarely, but in the presence of a tau mutation or traumatic brain injury or beta-amyloid deposition among other precipitants, tau is more likely to assume an aggregation-prone conformation. In vitro, polyanions (molecules or chemicals with negative charges) are capable of inducing tau self-assembly. Similarly, in living cells, contact with charged membrane phosphates or RNA may predispose tau toward an aggregation-prone conformation.[15] A compelling biophysical mechanism that may initiate the misfolding is the elimination of water that normally surrounds each tau molecule, and thereby makes tau sticky and prone to the aggregation into threads.[16] Once an oligomer or fibril is formed, it can seed subsequent reactions within the cell and enter neighboring cells through massive vesicles, termed macropinosomes.[17] Thus, when tau forms an aggregate, it also appears capable of transmissibility from cell to cell.

Patterns of tau spread constitute a neuroanatomical network, and these networks are associated with clinical features.[18] Given the described thinking concerning tau strains, the patterns of spread may arise from the specificity of a particular tau strain for a specific network. These selectively vulnerable patterns can be predicted by a diffusion mechanism modeled by a graph theoretic analysis using tractography data.[19] Neural network compromise may begin long before there is neuropathological evidence of disease in the form of misfolded tau aggregates. A study that recorded from neocortical pyramidal cells in a mouse model of tauopathy found numerous significant physiologic alterations when only a fraction of the neurons showed pathological tau.[20] Membrane potential oscillations were slower during slow-wave sleep and under anesthesia. Firing rates were reduced with longer latencies and interspike intervals. These changes reduce the activity of the neocortical network and suggest that conduction and synaptic transmission deficits may be among the earliest changes induced by tau spread at a resolution below light microscopy.

The interest level in tau among scientists has had the kinds of peaks and valleys that one might compare to the stock market. And like the last few years of the stock market, investment and the trajectory of growth have risen considerably. As we gain a deeper understanding of the molecule, as well as the ability to image tau pathology in the living human brain, we stand on the threshold of treatments.

6

New Movement in Neuroscience

A Purpose-Driven Life

By Adam Kaplin, M.D., Ph.D., and Laura Anzaldi

Adam Kaplin, M.D., Ph.D., graduated from Yale University before receiving his M.D. and Ph.D. from the Johns Hopkins University School of Medicine. Kaplin is the chief psychiatric consultant to the Johns Hopkins Multiple Sclerosis and Transverse Myelitis Centers and has a joint appointment as a clinician-researcher in the departments of psychiatry and neurology at Johns Hopkins, where his research focuses on immune-mediated mechanisms of depression and cognitive impairment in central nervous system autoimmune diseases. Kaplin is on the board of medical advisors to the Montel Williams MS Foundation, and a medical advisor to the Cody Unser First Step Foundation (CUFSF), the Transverse Myelitis Association (TMA), the Johns Hopkins Project RESTORE, and the Nancy Davis MS Foundation. He is the inventor and co-developer of www.mood247.com, a mood-tracking technology, and also the CEO and president of Altammune, a startup biotechnology company specializing in developing aggressive new therapies to put autoimmune diseases into long-term remission.

Laura Anzaldi is a rising third-year medical student at Johns Hopkins University School of Medicine. She graduated as the salutatorian from the University of Maryland, Baltimore County, in 2013, majoring in computer science, biology, and bioinformatics. She has not yet decided on a medical specialty, but her research interests focus on the intersection of technology and medicine. Anzaldi is particularly interested in developing software applications to enhance clinical workflow, facilitate research, and improve patient care and education.

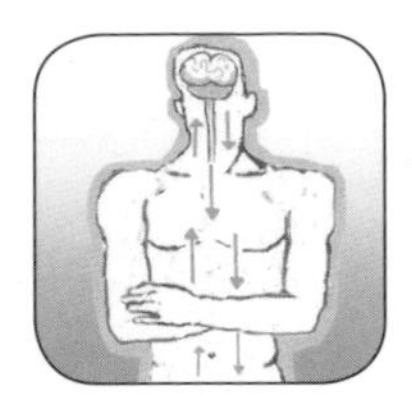

Editor's Note: Purpose in Life (PIL) is a research area that focuses on the interactions between mind and body and the powerful ways in which emotional, mental, social, and spiritual factors can directly affect health. It links the belief that your life has meaning and purpose to a robust and persistently improved physiological health outcome—particularly as a way to treat dementia, spinal cord injuries, stroke, and immunological and cardiovascular issues that include but extend beyond the brain. While it has inspired significant research, the authors contend that PIL is underappreciated, given its potential importance and interest to both the clinical and lay communities.

"He who has a Why to live for can bear almost any How." – Friedrich Nietzsche

WHY ARE WE HERE? What is the meaning of life? Existential questions such as these are captivating and considered fundamental to the human condition. Religions, philosophers, and scientists alike have sought answers for the human race as a whole, but the search for meaning also can be personal. People's perception of their own purpose may have profound consequences not only for the legacy they leave behind for others but also for the quality and quantity of their own life. We've all heard anecdotes of people who have suffered tragedies only to persevere with newfound purpose and zest for life.

These stories are certainly inspirational, but what if meaning also could soothe inflammation or protect neurons? What if finding purpose in your life could reduce your risk of dementia or stroke? That's the focus of research into what is now called Purpose in Life (PIL).

Interest in PIL, or the "mind-body axis," has ancient roots, but the study of it in individuals has attracted medical researchers' attention only recently. Current research reveals exciting correlations between PIL and positive health outcomes in a multitude of body systems. In the 1940s, Viktor Frankl introduced PIL to psychiatry. That Frankl was able to share his theory at all is nothing short of miraculous. He was a Jewish physician trained in both psychiatry and neurology who practiced in Austria when it came to be occupied by Nazi Germany. He survived three brutal years in various concentration camps, among them Auschwitz. He writes about his experiences in his magnum opus, *A Man's Search for Meaning*, where he also summarizes "logotherapy," a set of ideas that sustained him during the Holocaust and crowned his professional career.

As Frankl writes, "Man's main concern is not to gain pleasure or to avoid pain but rather to see a meaning in his life. That is why man is even ready to suffer, on the condition, to be sure, that his suffering has meaning."[1] Frankl emphasizes that this meaning is individual rather than general—people have to determine for themselves their mission in life. Compared to other psychologic doctrines that focus on looking back to the impact of past events, or inwardly through introspection, logotherapy looks to the future and to a person's will to do something meaningful with it.

As modern psychiatry began to evolve, the application of logotherapy to the treatment of psychiatric disorders, particularly those stemming from an "existential vacuum," was met by some with skepticism. The main criticism was that a person's perception of purpose had not yet been operationalized, measured quantitatively, or studied systematically. In an attempt to address this, two early investigators, James Crumbaugh and Leonard Maholick, created a psychometric scale in 1964 to assess PIL.[2] After scouring literature related to existentialism and logotherapy, they developed a 20-question PIL scale. Their scale and its derivatives have since been used in various populations and studies as a metric for PIL.

Other researchers have sought to characterize the nuances of PIL. The general consensus is that PIL includes dimensions such as (1) believing that life has meaning or purpose, (2) upholding a personal value system, and (3) having the motivation and ability to achieve future goals and overcome future challenges. PIL is a philosophical concept, but that has not stopped scientists from exploring its practical, biological impact. In particular, it seems that having a sense that one's life has purpose significantly supports the health of the central nervous system (CNS).

Protecting Cognitive Reserve

"Dementia" describes a global constellation of symptoms: memory, cognition, and communication problems. It commonly affects older people (most often age 60 and up), but it is not a normal part of aging. Dementia occurs when brain cells, or neurons, are damaged and no longer network properly. Different types of dementia are characterized by how and where the cell damage occurs.

The most common form of dementia, Alzheimer's disease, accounts for 60 to 80 percent of cases[3] and is the focus of much PIL research. Among the neurons most affected in Alzheimer's are those found in the hippocampus, a seahorse-shaped region of the brain associated with short-term memory. Through mechanisms still being elucidated, proteins called beta-amyloid and tau accumulate in neurons and lead to cell death and improper functioning. The damage in Alzheimer's primarily manifests as memory loss, starting with recent events and then more remote experiences.

The huge personal and public health implications of Alzheimer's have generated significant interest into ways to halt or prevent this illness. Other than generic advice to eat healthy, exercise, and engage in intellectually stimulating activities, researchers cannot yet tout strategies to reduce Alzheimer's risk significantly. However, new work by Patricia Boyle and colleagues at the Rush Alzheimer's Disease Center suggests that PIL could be neuroprotective (brain-preserving). After following more than 900 older people at risk for dementia for seven years, they found that those with a high PIL were only half as likely to develop Alzheimer's disease than those with a low PIL,[4] even after controlling for demographics, depressive symptoms, personality vulnerabilities, social network size, and number of chronic medical conditions. Those studied were also 30 percent less likely to develop mild cognitive impairment, a condition characterized by minor cognitive deficits that could (but doesn't always) progress to Alzheimer's.[4]

Boyle's group further explored the relationship between PIL and cognitive change over time. For people without Alzheimer's disease, a high sense of purpose was associated with slower rates of age-related cognitive decline.[4,5] In another experiment, Boyle's group looked at autopsy specimens of people who had been diagnosed with Alzheimer's and examined the amount of beta-amyloid and tau deposits in their brains. People who had a high PIL before death demonstrated better cognitive function, even in the presence of higher burdens of Alzheimer's-related protein accumulation.[5]

These studies suggest that PIL may have a protective effect on what is known as cognitive reserve. Researchers believe that people with more cognitive resilience ("cognitive reserve") at baseline are able to withstand more brain injury before developing neurologic symptoms. While the biological mechanism of this relationship is uncertain, it warrants more research.

The Heart of the Matter

Although the heart and blood vessels are not technically components of the nervous system, the brain and the CNS are inextricably linked to cardiovascular function. The heart's activity is intimately monitored and regulated by the brain, such as when your heart races when you are anxious or excited. When you experience these emotions, the brain initiates a series of events that lead to the secretion of adrenaline, causing the heart to beat faster. As William Harvey, the great pathophysiologist and father of investigations into the cardiovascular system, once said, "Every affectation of the mind that is attended with either pain or pleasure, hope or fear, is the cause of an agitation whose influence extends to the heart."[6]

Nor can the brain function without the heart delivering a reliable and timely supply of oxygen-rich blood. This delivery depends on vascular health. A stroke occurs when blood vessels fail to oxygenate brain tissue, whether because of hemorrhage or obstruction. Strokes may range from brief, reversible transient ischemic attacks to massive, deadly infarcts (brain tissue death stemming from a prolonged lack of oxygen and blood). Survivors can experience physical disability, such as paralysis, stiffness, dizziness, and fatigue, and/or higher cognitive disability, such as changes in mood, judgment, personality, or speech. Strokes are the fifth leading cause of death across the lifespan and a major cause of disability in the United States.[7]

While a healthy diet and regular physical activity are ways to reduce the risk of stroke, research suggests that having a sense of purpose also may play a role in prevention and prognosis. In one study, Eric Kim and his team assessed the level of PIL at baseline in almost 7,000 older adults who had never had a stroke and followed them over a four-year period to determine stroke incidence.[8] They found that for each standard-deviation increase in PIL score, these adults reduced their stroke risk by 22 percent.[8] The association held even after they controlled for behavioral, biological, psychological, and sociodemographic factors.

Other studies have looked at PIL and the risk of heart attacks. After following 1,500 individuals with cardiovascular disease for two years, researchers found that a higher baseline PIL was linked to a lower risk of heart attack. Each unit increase of baseline PIL (on a six-point scale) was associated with a 27 percent decreased risk of having a heart attack within two years.[9] Once again, these findings were still statistically significant after the researchers controlled for behavioral, biological, psychologic, and sociodemographic factors.

Researchers also have looked at the association between PIL and mortality, particularly from cerebrovascular causes. One study found that a strong sense of purpose was associated with a 72 percent lower rate of death from stroke, a 44 percent lower rate of death from cardiovascular disease, and a 48 percent lower rate of death from any cause in a population of men after an average of 13 years of follow-up.[10] This relationship held even after researchers controlled for perceived stress and cerebrovascular risk factors.

Reducing Inflammation

Inflammation has been implicated in many diseases that afflict the brain and nerves: from autoimmune CNS diseases (such as multiple sclerosis) to neurodegenerative diseases that share high rates of cognitive impairment and depression (such as Alzheimer's and Parkinson's). Inflammation is caused by activity of the immune system, which is made up of many cells and chemical mediators (called cytokines) that allow for communication between immune cells. Although inflammation is critical for clearing infection and healing wounds, excessive or persistent inflammation can lead to tissue damage and disease. Inappropriate immune system activity is thought to contribute to serious CNS maladies such as stroke, epilepsy, traumatic brain injury, Parkinson's disease, multiple sclerosis, and Alzheimer's disease.[11]

A less obvious contributor to inappropriate immune system activity is psychosocial stress. Our body's response to stress is controlled, in part, by what is called the hypothalamus-pituitary-adrenal (HPA) axis. Psychosocial distress is communicated by the brain (specifically the hypothalamus and pituitary) to the adrenal glands, which are located on top of the kidneys. The adrenal glands respond by secreting a stress hormone called cortisol. In the short term, cortisol actually depresses the immune system. However, when we experience stress for prolonged periods of time, the immune system stops responding as sensitively to cortisol. The result: more immune system activity instead of less. Although the evolutionary basis for this phenomenon remains to be fully elucidated, scientists believe that it is the body's way of gearing up for potential injury or infection related to the stressor.[12] This leads to sustained low-grade inflammation and higher levels of pro-inflammatory cytokines, which paradoxically can aggravate or cause disease.

One example of PIL and a link to positive, objective changes in inflammatory response is interleukin-6 (IL-6), a cytokine that is important in the proinflammatory initial response of the immune system to a host of general stimuli, including bacterial and viral exposure. IL-6 is one of the mediators that lead to the activation of the HPA axis and subsequent cortisol release. Dysregulation of IL-6 has been implicated in multiple CNS diseases, including cerebrovascular and Alzheimer's diseases. In an experiment that looked at the blood levels of IL-6 and its receptor in a population of women, researchers found that higher PIL scores were associated with lower levels of the IL-6 receptor, which implies less IL-6 activity.[13] This relationship held when researchers controlled for sociodemographic and health factors, and it suggests that PIL may be associated with a chronic calming effect on immune system activity.

Other studies have examined the impact of PIL on the inflammatory stress response. Lower levels of PIL are associated with increased sensitivity of the immune system, specifically IL-6, with repeated stress.[12] In other words, higher levels of IL-6 were detected in the bloodstream of participants with low PIL scores in each subsequent stressful exposure. Another study looked more generally at stress-related "transcriptomes." A transcriptome is essentially a collection of all of the genes that are expressed in a specific system. In this case, the researchers explored which genes were active in immune cells in people with hedonic or eudaimonic well-being. As defined by philosophers, "hedonic" well-being represents the sum of the positive emotional experiences that an individual has experienced, and "eudaimonic" well-being results from an individual's striving toward meaning and a purpose beyond self-gratification. Immune cells in people with hedonic well-being expressed more pro-inflammatory genes than did those in people with eudaimonic well-being.[14] This correlation implies that seeking purpose helps avoid a pro-inflammatory state, a positive step in fighting neurological diseases.

The Pursuit of Happiness

It's much easier to understand something tangible like a physical disease or medical treatment than something conceptual like purpose in life, and only in the past decade have researchers explored the connection between PIL and neurologic disease. Indeed, the pursuit of happiness receives a lot more attention in our culture than the pursuit of meaning or purpose. People strive for happiness, which is even considered a fundamental, inalienable human right according to the Declaration of Independence, and who could blame them? We feel happy when things go our way, and lower levels of stress and worry often accompany that feeling, at least briefly. It's a feeling rooted in nature: Even animals experience a sort of happiness when their needs are satisfied.[15]

But to derive meaning and thus identify a purpose in life is uniquely human and requires self-reflection and evaluation. Although both happiness and meaning play into overall life satisfaction, it may be possible to have a happy life without meaning or a meaningful life without happiness. A purely happy person is primarily concerned with the present and instant gratification of their own needs.[15] A person who pursues a chiefly meaningful life is more likely to contemplate the past or future and be concerned about others' well-being.[15] Meaningfulness is more enduring than happiness and can sustain people through periods of stress and suffering, as Frankl observed in the concentration camps.[15] Man's desire to find a purpose in life may even have played a crucial role in our development as a species, when we needed to band together against predators and the elements to survive.

The antithesis of happiness is depression. Depression is a disorder of mood characterized by persistent feelings of sadness, hopelessness, guilt, and apathy. Even Crumbaugh and Maholick's original paper commented on the apparent overlap between PIL and depression: "The tendency of highly depressed patients to show a loss of life purpose and meaning is clearly observable in the clinic."[2] People who are depressed have transiently lower PIL scores than people who are not, though it may be difficult to untangle whether depression decreases PIL or low PIL leads to depression.[16] But this difficulty does not muddy the correlation between PIL and health outcomes, as all of the studies that controlled for depressive symptoms still saw significant relationships. Depression is not the reason that people with low PIL have worse outcomes compared to those with high PIL.

What do we know about the relationship between high PIL and depression? Unfortunately, it seems that a strong PIL does not protect the very old from developing depression over a five-year period.[16] However, another group of researchers looked at teenagers, who, like the very old, are prone to depression. Instead of specifically looking at PIL, this research group explored the impact of hedonic vs. eudaimonic well-being on the development of depression. Research has shown that teens who were more eudaimonic (striving toward meaning and a purpose beyond self-gratification) had lower rates of depression one year later compared to those with hedonic well-being.[17] So in addition to improving nervous system disease outcomes in older people, meaningful and purposeful activities may improve the mental health of younger populations.

Additionally, we can draw some parallels between meaningfulness and peaceful feelings that religion can bring. Many people experiencing a tragedy or crisis turn to faith to find comfort, support, and answers. It is possible to endure almost anything as long as we can identify a greater purpose, and for some, religious doctrines and beliefs provide reasons and reassurances for suffering. However, research suggests a complicated relationship between one's religious beliefs and PIL, one that differs depending on how clearly and confidently individuals hold to their self-concepts of the world and their place in it. Researchers have shown that if a person has a low self-concept (for example, "My beliefs about myself often conflict with one another"), religion can return their PIL to baseline.[18] For those with a relatively high level of self-concept (for example, who endorse statements such as "I have a clear sense of who I am and what I am"), their level of PIL was unaffected by their degree of religiosity, demonstrating that PIL and religion are separate and independent phenomena.[18] This is consistent with the fact that PIL is self-defined and therefore subjective, and that the PIL metrics do not ask any specific questions about religiosity.

The Millennial Generation and Beyond

Our current societal climate is particularly primed to embrace PIL. The Millennial generation is just coming into its own, and its members may not be as entitled and narcissistic as they are commonly portrayed. A study by the career advisory board at DeVry University looked at Millennials' attitudes about employment issues, based on input from hiring managers. The study found that 71 percent of Millennials ranked finding work that is meaningful as one of the top three factors determining their career success, and 30 percent of Millennials ranked it as the most important factor.[19] Millennials are willing to make less money and work longer, nontraditional hours, as long as their work is personally meaningful.

This newfound cultural emphasis on meaning should revitalize research into PIL. While research has suggested significant relationships between PIL and positive health outcomes, we cannot yet make any sweeping declarations about PIL being responsible for those outcomes. This is primarily because PIL studies that prove causation are difficult to design. But scientists can explore other aspects of PIL, such as its natural history. Is PIL a constantly shifting quality that changes throughout life? On what time scale? One recent study in a very old population demonstrated that PIL decreases over a five-year period, especially in women and/or people with depression.[16] But the researchers looked at PIL only at the beginning and the end of the study period. It could be insightful to see how PIL changes on a more regular basis and how that relates to health outcomes.

Another interesting avenue is to identify specific interventions to increase someone's purpose in life. A high PIL has been linked to the pursuit of community-oriented goals, as well as to higher levels of physical activity.[20,21] Although the directionality of these relationships cannot be determined from studies thus far, they are important jumping-off points for future research.

Pharmaceutical treatments for any ailment that affects our minds and bodies absolutely have their place in healing, but they also can include significant potential side effects. Physicians should consider whether they are too quick to be pill pushers when they could be PIL promoters. Identifying a purpose to life can have profound implications in overall life satisfaction and health, as it motivates and drives us even in the face of difficulties and hardships. PIL appears to be biologically wired into our thinking and necessary for optimal health; a feature of our brain that defines each of us individually and simultaneously is a unique characteristic of the human condition.

7

Schizophrenia

Hope on the Horizon

By Patrick F. Sullivan, M.D., FRANZCP

Patrick F. Sullivan, M.D., FRANZCP, is a psychiatric geneticist, founding member, and lead principal investigator of the Psychiatric Genomics Consortium (PGC), and a Distinguished professor in the Departments of Genetics and Psychiatry at the University of North Carolina, Chapel Hill. He is also a professor in the Department of Medical Epidemiology and Biostatistics at the Karolinska Institute in Stockholm, Sweden. A native of St. Paul, Minnesota, Sullivan majored in biology at the University of Notre Dame and received his M.D. from the UC, San Francisco. He completed a residency in psychiatry at Western Psychiatric Institute and clinic at the University of Pittsburgh. He did additional training in psychological medicine in Christchurch, New Zealand, and is a fellow of the Royal Australian and New Zealand College of Psychiatrists.

Editor's Note: In July 2014, an international consortium of schizophrenia researchers co-founded by the author mounted the largest biological experiment in the history of psychiatry and found 80 new regions in the genome associated with the illness. With many more avenues for exploring the biological underpinnings of schizophrenia now available to neuroscientists, hope may be on the way for the estimated 2.4 million Americans and 1 in 100 people worldwide affected by the illness, one in which drugs have limited impact and there is no known cure.

WHAT IS MADNESS?

This is an extremely old question, one that has bedeviled generations of physicians and natural philosophers. Written documents identify schizophrenia (the more modern and precise term for madness) as mania, and it can be traced back to pharaonic Egypt, which dates to the second millennium BC. Clinical impressions and intellectual speculation have long dominated much of the discourse.

In the past two years, we have made considerable progress in understanding the fundamental nature of schizophrenia, a descriptor first coined by Swiss psychiatrist Eugen Bleuler in 1911. Modern genetics have provided these new insights.

Even so, schizophrenia continues to be a conundrum. On one hand, we know that it is a major cause of morbidity, mortality, and personal and societal cost.[1] For instance, schizophrenia ranks among the top ten medical conditions that cause significant, often lifelong, impairment and disability. The life expectancy of people with schizophrenia is around a decade less than of those who do not suffer from the disease, and the cost to treat patients is around $1.4 million over a lifetime. While a class of medicines can treat the major signs and symptoms effectively, particularly hallucinations and delusions, we also know that their benefits are incomplete and inconsistent. For far too many people, these medicines work only partially or, occasionally, not at all.

On the other hand, in spite of schizophrenia's status as a major public health problem, we have few hard facts about its fundamental causes. This lack of knowledge is not the result of a lack of studies. Schizophrenia has been studied intensively by several generations of scientists using the best methods and technologies available to them at those times.

But schizophrenia always has been particularly recalcitrant to scientific inquiry. Worse, research occasionally has attracted negative notoriety since the disorder was first wrongly identified as a form of dementia by Emil Kraepelin, M.D., in 1887. The field seemed to have a recurring pattern wherein one study would report a seemingly game-changing biological finding. The work would receive considerable attention in the media and engender excitement. But, repeatedly, other scientists could not replicate the original results.

For instance, about a half-century ago, it was reported that people with schizophrenia had a "pink spot" observed after a special type of urinalysis, suggesting the presence of an abnormal blood metabolite. After a lot of excitement and high-profile papers, researchers determined that the pink spot merely was a nonspecific lifestyle difference. Another example was the observation that psychotic symptoms improved after hemodialysis (suggesting a blood-borne toxin). Unfortunately, subsequent double-blind studies could not replicate initial findings. Genetics was no exception. In the past 25 years, perhaps a dozen genetic associations with schizophrenia received considerable attention but have not withstood the test of time and replication.

No Biology, No Mechanism, No Drug Targets

Thus, the origins of madness are elusive. The lack of fundamental knowledge has a major consequence: If we do not know the biological processes fundamental to schizophrenia, we can only make guesses about how to design better therapeutic and intervention strategies. But encouraging results from recent research have us entering a different and more productive phase in schizophrenia research.

To date, the development of antipsychotic medications—the backbone of treatment for many people with schizophrenia—has been serendipitous (the first in the class, chlorpromazine, was developed as a surgical premedication in the 1950s). Most subsequent approved antipsychotics were just variants on a theme (add a chlorine here, remove a methyl group there).

There was a boom ten to 15 years ago in the development of these medications; at the time, about one-third of the best-selling medications (mostly "me-too" chemical variations on other drugs) were antipsychotics or antidepressants. On one list, three of the top ten best-selling medications in history were antipsychotics. But following some failures and patent expirations, the antipsychotic drug development pipeline went dry.

Many believe that the lack of new and better medications for schizophrenia is the direct result of a lack of knowledge of its basic biology, which we need not only for medications but also to understand exactly who is at risk, to understand the impact of nondrug therapies (e.g., various cognitive remediation strategies), and to pick up early changes in patients that herald full-blown clinical worsening.

Why Genetics?

Many studies have involved family history, adoption, and twins. Taken together, these studies tell us that schizophrenia has a definite tendency to run in families and that it does so mostly via inheritance.

We should emphasize the word "tendency." The risk of schizophrenia to relatives of an individual with the disease is about tenfold. But because the lifetime risk of schizophrenia in the overall population is around 1 percent, only about 10 percent of the relatives of a person with schizophrenia will become ill—the majority of people with a relative with schizophrenia will not develop this illness. In fact, only a small proportion of people with schizophrenia have a relative with the illness.

So the genetic processes involved are probably subtle. This is markedly different from what people usually think when they hear "genetic." For example, in Huntington's disease, a neurodegenerative condition, the genetic signal is essentially deterministic: If you have the genetic risk variant, you almost certainly will contract the disease later in life.

The genetic risks for schizophrenia are more complex—probabilistic, not deterministic. This suggests that the biological processes underlying schizophrenia are relatively understated. It also is consistent with the clinical course of schizophrenia: a waxing and waning illness with preservation of basic neurological and higher cognitive functions and without an inexorable decline to dementia, intractable seizures, coma, and death.

Changing the Way We Do Business

Genetics is thus a major etiologic clue. In parallel, knowledge of the genome and the technologies to query the genome have made marked advances. It is now possible to do genetic assays with speed, throughput, and accuracy that would have been nearly inconceivable ten or 20 years ago.

Past impediments were human and structural. Typical approaches in psychiatric genetics were based on siloed research groups. There were a couple dozen research groups, and they tended not to work collaboratively. One would gather a sample and study it to the best of its ability. Time and time again other groups could not replicate the first one's conclusions.

We needed a new way to work. For true progress, we needed to find ways to cooperate in order to put together much larger samples than any single group could attain, and to introduce extremely high degrees of rigor.

A New Collaborative Spirit

In 2007, I was part of a group of psychiatrists, psychologists, and geneticists that founded the Psychiatric Genomics Consortium (PGC) on the principles of cooperation, democracy, and participation, and dedication to rapid progress and open sharing of results.[2] The PGC began with four National Institutes of Health (NIH) studies funded by the consortium's foundation. These studies were for schizophrenia, bipolar disorder, major depressive disorder, and attention deficit hyperactivity disorder (ADHD), and it seemed clear to us that we should try to become a trusted clearinghouse for conducting large and careful genetic analyses within and between these disorders.

The PGC has since become the largest consortium in the history of psychiatry, and arguably its most successful biological experiment. The PGC currently consists of more than 800 scientists from thirty-six countries. We currently have about 160,000 individuals in analysis and are in the process of adding about 153,000 more. We have published 17 major papers and 31 secondary-analysis or methods-development papers, and at least 75 other papers have made major use of our results. Schizophrenia is one of nine psychiatric disorders whose genetic basis PGC researchers are studying carefully.

Last July, the PGC published a landmark paper on schizophrenia in the journal *Nature*.[3] We reported the discovery of 128 different places in the genome where common genetic variation conferred risk for schizophrenia. This paper increased the number of intriguing genome locations by a factor of five. The paper had more than 300 authors and represented the concerted efforts of dozens of people for about three years.

The findings make a lot of sense. For example, one key genetic result was near the dopamine type 2 receptor gene. This finding represents a sort of biological convergence; this receptor is the site of action of virtually all affective antipsychotic medications. In addition, the findings implicated N-methyl-D-aspartate (NMDA) receptor subunits and neuronal calcium signaling, findings that also have independent lines of supporting evidence. These findings have allowed us to study the biology of schizophrenia with excellent starting points and tie together different types of genomic studies.

One important follow-up finding revealed overlapping genes for schizophrenia. Using modern technologies, we can study the genetics of schizophrenia in several different ways. The 128 findings above have significant overlap. This degree of convergence is remarkable and strongly suggests that we are circling in on the true nature of madness.[4,5]

What Causes Schizophrenia? What Is the Nature of Madness?

Genes clearly are involved in schizophrenia, but we have more hard work to do to tease out more specific answers than that. Although we have identified more than 100 places in the human genome involved in schizophrenia, there are hundreds more. Larger studies should help us develop a far more complete enumeration of the genes involved. These genes—what they do and when and where they do it—will deliver the biological clues needed to understand the mechanisms underlying schizophrenia.[6,7]

These are not the only important questions, either within or beyond genetics research. Among the additional challenges:

- Why do some people who have inherited a high risk remain well? Is it merely dumb luck or blind chance, or is there a method to the absence of madness? Someone can inherit a large number of schizophrenia risk factors and never develop schizophrenia. Future studies involving individuals at high risk who do not develop the condition could prove to be highly informative for understanding protective factors.

- Environmental factors are involved too. (Most geneticists are interested in the environment, but measuring genetic factors reliably is a far easier type of study.) The studies often cited for evidence that schizophrenia has a genetic basis also show that it has an important, albeit lesser, environmental basis. What is the biological impact of the key environmental factors? How do they act and interact with the genetic risks?

- What are the clinical implications of the emerging genetics knowledge of schizophrenia? Can we use this knowledge to improve diagnosis? Early on, can we single out individuals who are likely to have a severe course of illness and target them for more intensive treatment?

- Can we continue to improve the list of the stronger genetic causes of schizophrenia? We have a good start on such a list—genetic variants that increase risk of schizophrenia by as much as 20-fold. Are there others?

- Should all people with schizophrenia get a full genetic workup at first presentation? Can we identify rare individuals who actually have a different genetic disorder?

What's Coming

In the past two years, genetic studies have changed the landscape of schizophrenia research. After a lot of uncertainties and false starts, we now have a solid path to greater knowledge. Nothing is foolproof in science, but we now know what we need to do. We just need to do more of it. Thus, studies being conducted right now will add resolution and detail to the major advances from last year. Planned studies have the potential to add considerably more knowledge.

The PGC *Nature* paper from 2014 studied some 36,000 individuals with schizophrenia. The PGC is now in the process of increasing that number to around 60,000 people with this illness. Current plans include considerable expansion beyond that.

We don't now know the causes of schizophrenia, a disease that affects 24 million Americans and will afflict another 100,000 who will experience a first episode in 2015. The hunt for its causes is old, and schizophrenia has been unusually intractable to methods that have worked well for other diseases. But I and many of my colleagues perceive that we've turned a corner, that there are chinks in the thick armor surrounding schizophrenia's core biology. The deep hope is that we are now on a path that will make real differences on the near horizon.

8

The Holy Grail of Psychiatry

By Charles B. Nemeroff, M.D., Ph.D.

Charles B. Nemeroff, M.D., Ph.D., is the Leonard M. Miller Professor and chairman of the Department of Psychiatry and Behavioral Sciences, and director of the Center on Aging at the University of Miami School of Medicine. His research has focused on the pathophysiology of mood and anxiety disorders and the role of mood disorders as a risk factor for major medical disorders. He received his M.D. and Ph.D. degrees in neurobiology from the University of North Carolina (UNC) School of Medicine. After psychiatry residency training at UNC and Duke University, he held faculty positions at Duke and at Emory University before relocating to the University of Miami in 2009. He has served as president of the American College of Psychiatrists (ACP) and the American College of Neuropsychopharmacology. He has received the Kempf Fund Award for research development in psychobiological psychiatry; the Samuel Hibbs Award, Research Mentorship Award, Judd Marmor Award, and Vestermark Psychiatry Educator Award from the American Psychiatric Association (APA); and the Mood Disorders Award, Bowis Award, and Dean Award from the ACP. He is the co-editor-in-chief of the *Textbook of Psychopharmacology*, published by the APA. He is a member of the National Academy of Medicine.

Editor's Note: "Holy Grail" is a well-known metaphor for the eternal spiritual pursuit for truth and wisdom. It suggests that in order for us to find what no one has found, we must search where few have looked. In 2013, a group led by Helen Mayberg published a groundbreaking paper that sought an answer to one of the most discussed conundrums in psychiatry and neuroscience: Can specific patterns of brain activity indicate how a depressed person will respond to treatment with medication or psychotherapy? Our author examines the findings and discusses their potential impact on treatment for a public health problem that affects millions of people worldwide.

THE PERSONAL AND SOCIETAL TOLL of major depression is almost unfathomable. This year we will lose more than 42,000 people to suicide in the United States, the only top ten cause of death in this country that has increased year after year. Much of this tragic outcome can be attributed to untreated, poorly treated, treatment-resistant, and undiagnosed major depression. In this regard, it is worth remembering the familiar quote, attributed to Joseph Stalin, "A single death is a tragedy; a million deaths is a statistic." Perhaps the personal misery and tragedy of major depression are best exemplified by the award-winning novelist William Styron's personal account in *Darkness Visible*:

> What I had begun to discover is that, mysteriously and in ways that are totally remote from normal experience, the gray drizzle of horror induced by depression takes on the quality of physical pain. But it is not an immediately identifiable pain, like that of a broken limb. It may be more accurate to say that despair, owing to some evil trick played upon the sick brain by the inhabiting psyche, comes to resemble the diabolical discomfort of being imprisoned in a fiercely overheated room. And because no breeze stirs this caldron, because there is no escape from this smothering confinement, it is entirely natural that the victim begins to think ceaselessly of oblivion.[1]

I had the opportunity to get to know Styron well in his later years and can attest to the severity of his depressive symptoms—the absolute inability to experience pleasure of any kind and a feeling of hopelessness.

The consequences of untreated or unremitted depression are quite dire, including an increased risk not only for suicide but also for alcohol and substance abuse, as well as for a variety of major medical disorders (cancer, heart disease, stroke, kidney disease, and others). Perhaps of equal importance is the well-replicated observation that the longer a patient remains depressed, the less likely he or she is to achieve remission. Taken together, the linking together of findings indicates that the personal, societal, and economic consequences of undiagnosed or not well managed major depression are devastating and represent a major public health problem in the U.S. and worldwide. Indeed, the latest Global Burden of Disease study revealed major depression to represent a major cause of disability.[2] All of the aforementioned considerations serve as the major impetus for developing predictors of treatment response in depressed patients.

The Current State of Evidence-Based Treatments

The U.S. Food and Drug Administration (FDA) has approved around 30 antidepressant medications for the treatment of major depression. Among them are selective serotonin reuptake inhibitors (SSRIs). These drugs change the balance of serotonin in the brain, such as fluoxetine (Prozac), paroxetine (Paxil), sertraline (Zoloft), escitalopram (Lexapro), and citalopram (Celexa). Another family of medications, selective serotonin and norepinephrine reuptake inhibitors (SNRIs), help increase serotonin and norepinephrine levels in the brain, such as venlafaxine (Effexor), duloxetine (Cymbalta), and levomilnacipran (Fetzima). Still others in this family include bupropion (Wellbutrin), vortioxetine (Brintellix), mirtazapine (Remeron) vilazodone (Viibryd), nefazodone (Serzone), and trazodone (Desyrel). In addition, tricyclics and monoamine oxidase inhibitors, which are two classes of older antidepressants that work by inhibiting the brain's reuptake of serotonin and norepinephrine, are also approved but tend to cause more side effects than the other classes of antidepressants.

But pharmacotherapy isn't the only option; two other major classes of treatment are also available—psychotherapy and somatic nonpharmacological treatments. In randomized, controlled trials, cognitive-behavioral therapy (CBT) and interpersonal psychotherapy (IPT) repeatedly have been demonstrated to be effective in the treatment of major depressive disorder (MDD). Whether other forms of psychotherapy, such as insight-oriented, psychodynamically based therapy, are effective in major depression remains controversial. Brain stimulation therapies involve activating or touching the brain directly with electricity, magnets, or implants, and the FDA has approved three somatic nonpharmacological treatments for depression: electroconvulsive therapy (ECT), vagus nerve stimulation (VNS), and repetitive transcranial magnetic stimulation (rTMS).

ECT is generally considered the most effective of all depression treatments, although no head-to-head, randomized, controlled trial has compared it with other interventions. It generally requires inpatient hospitalization, at least initially, and general anesthesia with nine to 12 treatments over a three- to four-week period. Its cost and concerns about memory loss and the stigma associated with "shock treatment" has precluded its more widespread use. VNS and rTMS are both FDA-approved for treatment-resistant depression; the former requires an invasive surgical procedure. Researchers have conducted relatively few controlled studies of these devices compared with the vast number of pharmacotherapy and psychotherapy treatment trials.

A Personalized Approach

With a plethora of drugs and psychotherapy approaches available, let us consider the problem psychiatrists encounter on a daily basis. A patient in my own practice serves as an example. A 50-year-old academic physician suffers from a classic major depressive episode associated with severe work stress. He has difficulty falling asleep, awakens several times during the night, and rises early with severe anxiety. He has reduced appetite, difficulty concentrating, and trouble enjoying any leisure activities, and he feels pessimistic about the future. He admits to passive contemplations about suicide, with recurring thoughts that if a car jumped the median and landed on his car, it would be an end to his suffering. He has no prior episodes or family history of depression. He has no underlying medical disorder that might be contributing to depression, such as hypothyroidism or drug or alcohol abuse. What treatment should I recommend for him? Antidepressants, and if so, which one? Psychotherapy, and if so, which one? One of the somatic, nonpharmacological treatments?

I want to recommend the treatment most likely to be successful in producing a complete remission of his depressive syndrome and relieving him of his considerable misery. What are the known and best-validated predictors of response? Our group has previously reviewed the scientific findings in this area.[3-6] The most reliable predictor is past response, but in this case the patient has never been treated for depression. A positive response in first-degree family relatives is also predictive of a beneficial response to antidepressants, but again, this is not applicable to this patient. Some evidence suggests that certain subtypes of depression respond best to certain treatments—monoamine oxidase inhibitors (the first type of antidepressants developed) are believed to be the most effective for patients with so-called atypical depression characterized by hypersomnia, overeating, extreme rejection sensitivity, and feeling better in the morning than later in the day. Combinations of antidepressants and antipsychotics or ECT are best for patients with major depression with psychotic features. However, neither of these subtypes are relevant to the patient I have described. What then can guide my recommendations? Surely patient choice is an important consideration, but it will likely be guided by my discussion with the patient.

In 2013, Helen Mayberg and her colleagues published groundbreaking findings that help in this case. One important caveat and disclosure: Several of the authors (Drs. Mayberg, Holtzheimer, Dunlop, and Craighead) were colleagues of mine for many years, and we continue to collaborate on various projects. However, I was not involved in the following study.

Mayberg's study sought to identify a biomarker that could predict which type of treatment would benefit a patient based on the individual's brain activity. Using regional brain glucose metabolism as measured by positron emission tomography (PET) as a proxy for neural activity, her group sought to determine whether baseline resting state activity predicted remission after 12 weeks of treatment with either the selective serotonin reuptake inhibitor escitalopram (10 to 20 mg per day) or 16 sessions of cognitive-behavioral therapy. The study sample initially comprised 82 men and women who were randomized between the two treatments. Of these, sixty-five patients completed the study and thirty-eight had clear outcomes and acceptable PET data. The 38 patients who comprise the analyzable data set were distributed as follows: 11 who went into remission with escitalopram (six nonresponders) and 12 who did so with CBT (nine nonresponders). The major findings were that hypometabolism of glucose in the insula, likely reflecting reduced activity of neurons in this brain region, was associated with remission using CBT, and with poor response to escitalopram. Contrariwise, insula hypermetabolism, reflecting increased activity of neurons in this brain region, was associated with remission using escitalopram and with poor response to CBT.

The authors conclude that baseline insula metabolism is the first objective marker to guide initial treatment selection in depression. Closer scrutiny of their data is worthwhile. First, they eliminated from their primary analysis the responders to CBT or to escitalopram who did not go into remission. More specifically, partial responders to escitalopram or CBT were excluded from the analysis. They did so in order to accentuate the differences between the extremes in the depressed population; the results revealed clear differences in glucose metabolism in six regions: the right anterior insula, right motor cortex, left premotor cortex, right inferior temporal cortex, left amygdala, and precuneus.

When all six regions were compared, the right insula exhibited the greatest effect as a discriminator of treatment response, followed by the precuneus. When the whole sample was studied, right insular activity was positively correlated with the depression symptom severity scale, and with the Hamilton Depression Rating Scale (HRSD) score in the CBT treatment group while right insular activity was negatively correlated with the HRSD in the escitalopram treatment group.

This finding is quite provocative. If additional research can replicate these results, it suggests that a simple brain imaging test could reliably pre-

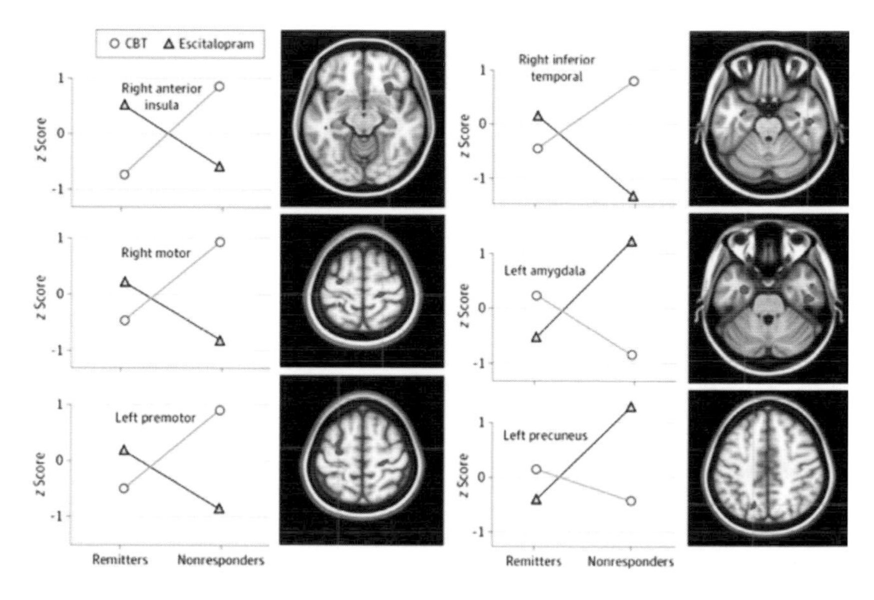

Potential Treatment-Specific Biomarker Candidates: Mean regional activity values for remitters and nonresponders segregated by treatment arm are plotted for the six regions showing a significant treatment × outcome analysis of variance interaction effect. Regional metabolic activity values are displayed as region/whole-brain metabolism converted to z scores. Regions match those shown in Table 2. Escitalopram was given as escitalopram oxalate. CBT indicates cognitive-behavioral therapy.[7]

dict whether a given patient should be treated with psychotherapy or antidepressant medication. It also raises a plethora of additional questions:

- A wealth of data, now summarized in a research meta-analysis, indicate that MDD patients with a history of child abuse and neglect exhibit a poorer response to pharmacotherapy and psychotherapy and exhibit unique brain imaging differences.[8] Mayberg's research does not address this critical clinical characteristic in this population.

- It is somewhat unclear how the six brain regions of interest were identified and why several regions repeatedly identified to be implicated in the pathophysiology of depression either were not selected or exhibited no significant effect, including the hippocampus, subgenual cingulate, and others.

- It is hard to know what to make of the findings that only the right anterior insula, right motor cortex, left premotor cortex, left amygdala, left precuneus, and right inferior temporal region show dramatic differences in the CBT versus escitalopram-induced remission versus nonresponder groups, whereas their counterparts—namely the left anterior insula, left motor cortex, right premotor cortex, right amygdala, right precuneus, and left inferior cortex did not. Was a composite of the left and right sides of these structures informative?

- As the authors themselves point out, the study comprises a relatively small number of patients and our field is replete with pilot study findings that, unfortunately, have not been replicated in larger trials.

- This study utilized PET instead of the more often used functional magnetic resonance imaging (fMRI) technology. As Mayberg and her colleagues appropriately point out in their paper, fMRI studies have examined regional brain activity and, more recently, resting state connectivity to identify MDD or MDD subtypes, but neither type of imaging has been used to discriminate response either among antidepressants or between antidepressants and psychotherapy.[9]

The expanding area of genetics in general, and pharmacogenetics in particular, is also of vital importance. A burgeoning database documents the role of certain genetic variations in vulnerability to mood disorders and, more recently, how variations may affect treatment response to different antidepressants. Whether genetic material was collected in Mayberg's study is unclear, but this focus is crucial, particularly in view of recent findings in imaging genomics. The lack of random assignment of the MDD patients as regards, for example, the vulnerability gene variants of the serotonin transporter or others, now shown to be associated with clear alterations in regional brain activity, could have confounded the results.

Such possibilities should not detract from the groundbreaking findings. This research group has always been willing to take great leaps forward, and they should be applauded for it. Subsequent studies will reveal if the insula is truly "the region" that predicts response to CBT versus a selective serotonin reuptake inhibitor, such as escitalopram, or whether other regions or biomarkers also need to be a component of the ultimate formula. This is part of the ongoing and exciting scientific process that is emblematic of the marriage of neuroscience and psychiatry. Ultimately, I believe this work will be judged as crucial in eventually attaining the goal all of us seek: a valid predictor of individual treatment response in depression, still the Holy Grail in psychiatry research.

9

No End in Sight

The Abuse of Prescription Narcotics

By Theodore J. Cicero, Ph.D.

Theodore J. Cicero, Ph.D., is the John P. Feighner Professor of Psychiatry at Washington University School of Medicine. After receiving his Ph.D. in 1968 from Purdue University, he joined Washington University's faculty as an assistant professor in 1970. Cicero has published more than 200 scientific articles in the general area of substance abuse research. Much of his work has focused on neurobiological substrates of dependence and abuse of drugs (opiates and alcohol) in animal models. Over the past 15 years, Cicero's focus has been on the epidemiology of opiate abuse, beginning with the emergence of the prescription drug problem in the 1990s and continuing with the recent transition of heroin abuse from an inner-city problem most common in poor, minority males to an epidemic in white, middle-class male and female residents of suburban and rural locations.

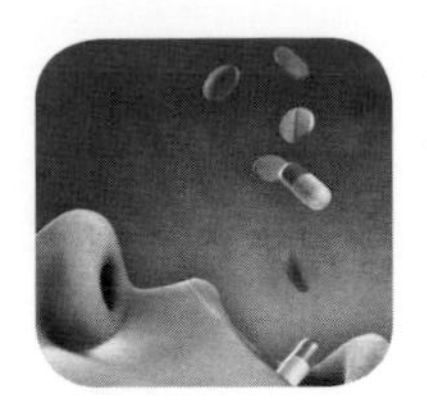

Editor's Note: From teenagers dying from heroin overdoses to crime tied to Vicodin and OxyContin addiction to road fatalities in which sedatives and muscle relaxants are involved, 20,000 deaths in the United States in 2014 were attributed to problems associated with narcotics and prescription drug use. Our author, whose research involves the neurobiological basis of drug addiction, traces the history and evolution of narcotics, and leans on his clinical experience to discuss why certain drugs are powerful, addicting—and dangerous.

FOR CENTURIES UPON CENTURIES, people have used opium or its components—morphine and other similar narcotics—to get "high" or feel mellow.[1,2] Opium is tied to ancient civilizations in Egypt, Persia, and China and to such figures as Alexander the Great, Hippocrates, John Keats, and John Jacob Astor. Paintings portray old Chinese men relaxing and apparently deep in thought in rocking chairs, smoking pipes filled with opium, dried latex that comes from poppies grown in many regions in Asia. In countries such as Afghanistan[3] and Myanmar, opium production and exportation is the basis of their economies.

While raw opium products historically have been highly prized mood-altering drugs, one of their main active components, morphine, converts very easily, with a simple chemical step, to heroin. Users appreciate that heroin travels to the brain much more quickly and effectively than morphine. Although once in the brain heroin breaks down to morphine, heroin is the drug of choice because its high is much quicker and more intense.

In the last 200 years, opium and its derivative, heroin, have also enjoyed enormous popularity for its potent pain-relieving (analgesic) effects. The unfortunate, often unrecognized, downside of opium and heroin is that both drugs are powerfully addictive, partially because they are snorted, smoked, or injected, which produce very intense and immediate effects. Of course, our concern today is not only for opium and its derivatives but also for the myriad of structurally related narcotic analgesics that have been developed in laboratories. Few would argue that opiate abuse is now a national crisis.

How Narcotics Work

All derivatives of the opium poppy are classified as narcotics and share common properties in the brain. All narcotics enter through the central nervous system (brain and spinal cord) to exert their effects. In the brain they bind to specific opioid "receptors" (so-called G-protein coupled receptors) on brain cells that use the electrochemical messenger GABA (gamma-aminobutyric acid) to communicate with one another.

There are at least three major subtypes of opioid receptors: mu, delta, and kappa. Each of these, particularly the mu receptor, have multiple subtypes (some estimates put the total number of opioid receptors at 25 to 30). Furthermore, while there is considerable overlap of receptor networks in the brain, each receptor is localized in a distinct brain region and has its own communication pathway. Why this diversity and extensive network in the brain? This is not totally known, but probably relates to the fact that "endogenous" opioids are produced by cells in the brain (as opposed to exogenous narcotic drug opioids) that regulate pleasure, pain, appetite, sexual behavior, hormonal balance, gastrointestinal activity, respiration, and other bodily functions via effects on discrete brain areas. So, while endogenous opioids play a fundamental role in many bodily functions, narcotic drugs similarly interfere with these same systems.

Pain relief produced by opioids is facilitated primarily by mu-opioid receptors and, to a lesser extent, by delta receptors. While endogenous opioids bind to these receptors to alleviate pain, such as that arising from stress, it is primarily narcotic drugs that bind to these receptors to relieve pain. All narcotic analgesics, arising internally or from narcotic drugs, have a high affinity for the mu-opioid receptor: the higher the affinity, the greater the response.

Narcotic analgesics work at the spinal cord level to overexcite nerve fibers so that they are poor conveyers of pain signals to the brain.[4] These drugs also exert powerful effects in the brainstem to both transmit signals to the spinal cord to dull the transmission of pain and to lessen the conscious perception of pain.

Unfortunately, the very property of certain narcotics that makes them excellent painkillers also makes them the most rewarding and addictive drugs known to humans. Indeed, all effective opioid analgesics produce euphoria and this effect is mediated almost exclusively by mu opioid receptors, just like pain. Thus, the correlation between the potency of analgesia and euphoria is nearly perfect. This explains why it has been literally impossible to develop a narcotic drug that is useful for pain but devoid of addictive properties—despite efforts over the last 100 years.

The ways in which opioids produce rewarding effects is complicated, but it appears that the neurotransmitter dopamine is intrinsically involved in the rewarding properties of all drugs, especially opioids.[5] Moreover, it has been possible to trace the reward pathway in some detail. The pathway involves perhaps a dozen or so different brain areas, several different neurotransmitters, and some stress-related factors.

It is important to note that euphoria, or rewarding pleasurable effects, are complex emotions, undoubtedly mediated by many regions in the brain. What we are trying to do is determine the molecular basis of a very elusive target: a thought or a feeling. Neuroscience has not yet reached the point where emotions can be reduced to biology.

Developing Addiction

While it is clear that many people who use opioids for pain management or to feel better handle their consumption well, others develop addictive patterns of use characterized by rapid tolerance, physical dependence, and, most importantly, craving. What happens in the brain to produce these effects? The brain adapts (develops tolerance) to the presence of high concentrations of drugs. The entire purpose of tolerance is to restore homeostasis, even in the presence of high levels of an environmental toxin, such as a drug. The downside, of course, is that more drug is required to produce the same effect, thus initiating further neuroadaptive changes. The mechanisms underlying tolerance are fairly well understood: It takes more drug to elicit a response from receptors, but the end result is that the brain is under constant assault and keeps adapting in ultimately non-productive ways to keep up with the insult produced by ever-increasing drug doses.

What happens, then, when the brain's neuroadaptations are confronted by abrupt cessation of drug use? A withdrawal response ensues, a hyperactive reaction that is generally the opposite of the effects produced by heavy use of the narcotic: Instead of drug-related constipation and elation, diarrhea and depression occur. While it was once thought that the mechanisms underlying withdrawal were confined to hyperexcitability after the neuroadaptations were disrupted, it now appears that other brain areas that were recruited during the development of dependence are affected as well, and responses by these areas may be just as—or even more—important than the direct result of the neuroadaptive responses. One thing we know for sure is that the withdrawal response is profound and often contributes to relapse.

While obtaining relief from withdrawal probably plays an important role in addictive processes, physical dependence and drug withdrawal occur with many drugs, yet most people do not develop the craving that is associated with addiction. Craving distinguishes physical dependence from addiction. So while clearly craving must be present for a diagnosis of substance-use disorder (SUD), some other mechanism beyond physical dependence also must be involved in craving. The nature of these other mechanisms is only now being examined.

From a neurobiological perspective, the answer to why some withdrawals are worse than others is largely unknown. But we do know that the perception of the depth of withdrawal symptoms and their unpleasantness is unique to certain drugs. With opiates, the difficulty of the withdrawal response is among the worst.

Government Steps In

In 1914, the Harrison Narcotics Tax Act made opium and opium products illegal unless prescribed by a physician.[6] Before that, many artists and intelligentsia used freely available opium to enhance their artistic experience, while others—such as Abraham Lincoln's wife, Mary Todd Lincoln—famously used it to escape depression and feel better about themselves.[7] Additionally, many elixirs were sold over-the-counter or by "snake oil" salesmen at carnivals and fairs, and their use was condoned. Instead of eliminating narcotics abuse, though, the Harrison Act perpetuated an extensive black market, an underground network that still exists.[8]

Shortly before World War II, all the way through the Reagan era, narcotics steadily gained momentum for pain relief, but abuse rates stayed relatively low.[9] When abuse occurred, heroin was mainly the drug of choice, particularly among inner-city male minorities seeking to feel better and escape their poverty and hopelessness. Despite drug wars and drug-related murders, mainstream society mostly ignored the heroin problem because the problem occurred in a marginalized segment of the population.

In the year 2000, a huge momentum shift came about when a nationally recognized non-profit health standards-setting and accrediting body, the Joint Commission on the Accreditation of Healthcare Organizations (JCAHO),[10] released a scathing report on the undertreatment of pain. It concluded that effective narcotic analgesics were available but seldom used, and that doctors were ignoring pain management because of an irrational fear of addiction. They argued that narcotics should be more widely used, since an anecdotal report in a prestigious medical journal found that few patients abused their narcotic medications.[11] As referenced in an Institute of Medicine Committee report in 2011, the JCAHO report reasoned that pain should be the fifth vital sign, meaning that doctors should routinely ask about pain as part of any physical exam, not wait until the patient volunteers this information, and treat it accordingly. This report made headlines nationwide; *Time* magazine featured it on its cover.

The report and subsequent campaign became successful to the point that doctors began prescribing narcotics in record numbers, some probably inappropriately.[12] Inevitably, with that many new prescription drugs now in medicine cabinets and on nightstands, quantities were diverted by people who sought not pain relief but a high (or a profit on selling the drugs for such purposes). Prescription narcotic abuse quickly reached epidemic levels, fueled by drug companies that rushed to meet the new demand.

Oxycodone Arrives

New products began arriving on a fairly regular basis in the 1990s, but the most impactful of these was a novel type of product: a sustained-release drug, oxycodone, that would provide pain relief for eight to 12 hours.[13] Oxycodone is an opioid agonist with very high affinity for the mu opioid, making it an excellent pain reliever but also a powerful euphorogenic agent. The drug—marketed as OxyContin—was attractive because it needed to be taken only once or twice a day, instead of every two to four hours. The long-lasting relief was particularly beneficial for older patients who suffered from memory loss and for people with limited mobility.

The extended-release capsules work by containing copious amounts of drug that a built-in delivery device would release slowly over time. Given its slow-release properties, the Food and Drug Administration (FDA) concluded that the delay in reinforcement would dissuade abuse because addicts typically seek an immediate reward. Thus the FDA, now infamously, allowed the sponsoring company to state in the package insert or label that abuse was expected to be low.

This was an ill-informed blunder of epic proportions. What the manufacturer claimed it did not recognize, nor apparently did the FDA, was that addicts cleverly and quickly realized that they could defeat the slow-release devise by crushing pills and making large amounts of oxycodone immediately available in a form suitable for snorting or intravenous injection—an immediate rush akin to that of intravenous heroin.

The ease with which addicts could breach the slow-delivery device—with Internet "how-to" tips posted within days—makes it somewhat difficult to believe that the manufacturer and the FDA were not quickly aware of their collective blunder and did not take aggressive action to rectify a very bad design. The pharmaceutical company's lethargic response to the revised FDA mandates may explain the $635 million in fines they received for marketing strategies that failed to recognize or mention the potential for abuse.[14]

Developing Addiction

The quality of the rush or high that addicts seek, particularly after the intravenous injection of narcotics, is characterized in many ways—including, graphically, as a whole-body orgasm. This very powerful sensation is a huge part of what keeps people dependent. Why these drugs are so rewarding is elusive. What is probable is that some individuals are genetically predisposed to have a much more pleasurable response than others. Still, the precise nature of this predisposition is unknown. Several quotes from patients in our clinical research program illustrate the powerful sensation these drugs elicit and the constant yearning that ensues:[15-17]

> "I found a bottle of 5mg hydrocodone tablets my dad had after knee surgery, and I took 1½ tablets. I was 18 years old and had just started smoking pot and experimenting with drugs. I went to bed, and after about an hour I felt an intense warm, fuzzy, pulsating euphoria come over my entire body. It was pure bliss and felt extremely good. I immediately always sought out opiates before any other drugs after that."
>
> "The high or reaction to it was the first of its kind, tingly body and feeling of being in a cloud. It lasted longer than I expected. I never achieved the same feeling from them again however I continuously searched for it."

The most logical question about the surge in the abuse of prescription narcotics is why it took off in such dramatic fashion, given that narcotic analgesics and heroin had been available for years. For starters, there was a seemingly endless supply of narcotic drugs, given their widespread therapeutic use and the proliferation of "pill mills" (pain clinics that carelessly dispensed huge amounts of narcotics for profit) and disreputable "script" doctors who would write prescriptions, whether they were needed or not, for quick cash.

Second, what makes prescription narcotics acceptable in many users' minds is that they produce a good, dependable, "safe" high. Unlike heroin, the dose is known with certainty, the pill labeled clearly, and they are legal (the latter often is not true, depending on extenuating factors). To justify their use and assuage any guilt, patients also may tell themselves, "At least it's not heroin; I'm not a heroin addict."

Regarding heroin, at least, they may have a point: Heroin bought on the street is usually sold in nonsterile powder form and is rarely more than a few percent pure, and some of the powder additives (talcum powder, quinine, sugar, and sometimes other drugs, including powerful opiates such as fentanyl) that make up the rest can be very dangerous, particularly in those in whom the IV route is employed. Of course, the IV route also introduces the possibility of blood-borne pathogen transmission and, given the uncertainty of the purity of street-purchased heroin, there is a distinct possibility of overdose. Thus, for most narcotic drug users who administer their drugs orally, the use of heroin, at least in the beginning stages of abuse, was considered to be a degenerate, unsafe option.

However, the climate for heroin and prescription drug abuse has changed. Most new users are not impoverished minorities from inner cities, but middle-class men and women living in suburban and rural areas who find it easy to justify their use. Thus, a perfect storm has developed: legal narcotics readily available with little or no social stigma attached to their use.[18]

The Great Escape

Other than the obvious high, what purpose do these drugs serve that accounts for their popularity? It turns out that the initial potent high is not really what most users seek. Rather, narcotics relieve anxiety or depression by providing a short-lived escape from difficult circumstances. For those who become addicted, the initial high is pure bliss and something they continue to seek, often for years. But pure bliss becomes an elusive goal and does not repair emotional dysfunction and unpleasant circumstances. More often than not, the user's life gradually disintegrates with addiction.

Several quotes from our clinical study participants are excellent representations of the utility users often find with narcotics, beyond their initial reaction:[15,16]

> "I didn't know who I was anymore, I didn't like who I was and didn't want to be in my own skin. I would use to feel good about myself, feel comfortable, confident, beautiful. Also, just life things in general, the unknown, life is scary and using was my escape to not deal with responsibilities."
>
> "I used drugs to hide the pain of not loving/accepting myself in the beginning. Then my father passed away and I used drugs to mask the pain of grieving for many years. My life quickly spiraled out of control and I used drugs to hide how bad things really were."
>
> "It numbs pain, emotional and physical. It makes everything temporarily seem amazing, even when everything around you sucks."

Beginning in 2010, prescription narcotic abuse began to level off,[19,20] and heroin began to resurface.[21-23] The primary difference with this current surge is that it migrated from primarily an inner-city problem to the same type of users who became the norm for prescription narcotic abuse.[21-23] The reasons—as comments from our clinical research program indicate[21-23]—are surprisingly straightforward. For example:

> "It [heroin] was most cost effective in terms of the high established over the more expensive and less effective [O]xy[C]ontin, Percocet, [V]icod[i]n, etc."
>
> "You could get more for the price. Around my area 'Southern Oregon' one [OxyContin] 80 mg (bluish green pill) goes for 70-80 bucks nowadays; used to be 50 but they are hard to find now because of changes made by the pill companies to make the [Oxy] unshootable, unsnortable and unsmokable. Heroin is always coming up from Cali … and is in high demand around here."

In addition, and perhaps as or more important, the social stigma associated with heroin use began to dissipate:

> "I knew I liked it [heroin] above all else, and once I had a drug dealer it became almost too easy to get, I had access to money because I am an upper middle class family and I also became close to my dealers, driving them around so I could get paid in drugs and just becoming super close, even if it meant sexually, so I could get the drug. The two

> dealers and the people around them that I developed that relationship with are also middle class white kids, not even kids we were all in the age range of 25-41. It just became easy, and we weren't really looked at as being addicts because everyone thinks heroin addicts are all homeless, shady looking, dirty junkies."

What's next? Where do we go from here? It seems obvious that while we could and should do all that we can to reduce the supplies of heroin and its legal counterparts, we must also reduce the demand. The evidence shows, as outlined in part above, that narcotics satisfy—in a very harmful way—a variety of "needs" in individuals who use them. We need to find better ways to persuade users that narcotics are not the answer. There have to be, and are, better ways than entering the vicious cycle of abuse to meet those needs.

Until we make a serious effort as a society to implement those alternatives, expect to read and hear more about the devastating impact, including overdose deaths, of the continued use of legal and illegal narcotic compounds. What factors, if any, will break our centuries-old love affair with opium and other narcotics? The drug formulations may change over time, but the appetite for narcotics has persisted for a very long time, with no obvious end in sight.

10

The Binge and the Brain

By Alice V. Ely, Ph.D., and Anne Cusack, Psy.D.

Alice V. Ely, Ph.D., is a postdoctoral research fellow in biological psychiatry and neuroscience at the UC, San Diego's Eating Disorders Center for Treatment and Research. Ely's work has focused on behavioral and neurobiological risk factors for weight and eating disorders, primarily concerning frontostriatal reward processing in response to anticipation and receipt of food reward. Her specific focus is on impulsive decision making in individuals with bulimia nervosa and the relationship between neurobiological response to food reward and anxiety. She received her undergraduate degree in psychology from the University of Pennsylvania and her Ph.D. in clinical psychology from Drexel University, following her clinical internship at the UC, San Diego and the San Diego VA Health System.

Anne Cusack, Psy.D., is a postdoctoral fellow at the UC, San Diego Eating Disorders Center for Treatment and Research and co-manages the adult treatment program. Cusack's research focus has been on the development of novel treatments for eating disorders, including acceptance, mindfulness, emotion regulation, and self-injurious behaviors. Her clinical focus has been dialectical-behavior therapy for binge-eating disorder, as well as mood disorders, substance-use disorder, post traumatic stress disorder, and personality disorders. A graduate of the Chicago School of Professional Psychology, Cusack completed her predoctoral internship at Greystone Park Psychiatric Hospital, working with various psychiatric populations. She currently co-chairs the Dialectical Behavior Therapy special interest group for the Academy for Eating Disorders.

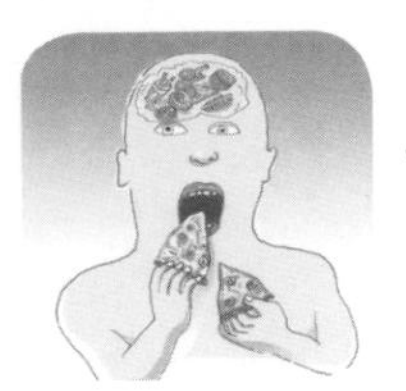

Editor's Note: Who hasn't dipped into that pint of Häagen-Dazs with the best of intentions and ended up finishing the entire container? Knowing where the line is when it comes to out-of-control impulse consumption is at the heart of binge-eating disorder (BED), a newly recognized mental condition that affects millions of people and that we are just beginning to better understand—from both a neurobiological and a clinical standpoint.

THE WORD "BINGE" IS USED CASUALLY to describe a meal that was larger than intended, or even a long evening of Netflix. But genuine binge eating looks very different from Thanksgiving dinner in which we go back for seconds on mashed potatoes, or a Super Bowl party where we devour an abundance of chips and dip. Genuine binge eating is recurrent and debilitating—physically and emotionally. Its medical consequences include high blood pressure, type 2 diabetes, high cholesterol, gallbladder disease, digestive problems, heart disease, and metabolic syndrome. It is frequently accompanied by anxiety and depression too, though these, like the binge-eating behavior itself, are relatively hidden.

Research suggests that more than 20 percent of college-aged women have engaged in binge-eating behavior, and that a similar percentage of the broader population will be affected by binge-eating disorder (BED) at some point in their lives.[1] Currently, about 60 percent of diagnosed cases are in women, and BED researchers believe that fewer than half of all cases receive adequate treatment. Indeed, research into the neural mechanisms of BED is still in its infancy, so there is no specific treatment for the disorder.

What Makes a Binge a Binge?

BED was added to the *Diagnostic and Statistical Manual of Mental Disorders* (DSM-5) only in the 2013 edition.[2] However, similar patterns of signs and symptoms have been reported in the medical literature since 1932, when German psychoanalyst Moshe Wulff described an eating disorder characterized by binge eating, depression, and disgust with one's body.[3] DSM-5 criteria for a BED diagnosis include at least one binge episode per week without any compensatory purging behavior (which would then lead to a bulimia nervosa diagnosis). Binges must involve an objectively large amount of food, consumed within two hours, and accompanied by a feeling of loss of control.

Loss of control is what distinguishes a binge from simple overeating, and similar constructs exist in substance use disorders. Loss of control can take many forms, but there are commonalities in patients included in the DSM-5 that are necessary criteria for a BED diagnosis: the inability to restrain oneself from consumption, a feeling of fullness, eating alone because of being embarrassed about how much one is eating, or even a sense that the food's taste is no longer appealing. Some also report eating much more rapidly than normal.

BED patients frequently report that their binge foods are foods that they view as forbidden or unhealthy. One of Wulff's patients, for example, described secretly eating large quantities of sweets, bread, and pastries because they were restricted from her daily diet due to medical obesity. "The worse [for me], the better," she explained.

In severe binges, the drive for overindulgence may cause patients to consume raw pancake batter, entire loaves of bread, frozen fish sticks, or other bulk foods, however lacking in taste they may be. The patient often plans secret eating in advance and carries it out late in the day—and typically experiences disgust, depression, and guilt afterward.

A patient in our clinic serves as an illustrative example. B, a 57-year-old African-American woman, reported significant struggles with food prior to entering treatment, and stated that she often thought about eating all day long. B said that she visited the grocery store several times a week, ate directly from bags or containers, and often consumed strange combinations of flavors, such as licorice, bread, peanuts, and beef jerky. Her remark that she "didn't even want those foods but couldn't stop eating them" pointed to a loss of control. B also reported frequently eating despite not being hungry, in order to fill a "void." She has tried to combat the resulting significant weight gain through a variety of fad diets, "wellness programs," and gastric bypass surgery. This has led to extreme weight fluctuations over 25 years, including a recent episode in which she lost 75 pounds on a crash diet. However, failing to address the psychopathology underlying her bingeing, she gained all that weight back.

People who meet the criteria for BED are more likely to be in the obese or overweight ranges, and to report struggling with their weight as children. Compared to healthy controls, those diagnosed with BED reported more frequent family histories of depression, greater vulnerability to obesity, more exposure to negative comments about shape or weight, greater perfectionism, and higher negative self-evaluation. Compared to obese individuals who do not binge, those diagnosed with BED reported greater weight and shape concerns, more personality disturbance, more mood/anxiety disorders, and a lower overall quality of life.

These factors highlight the importance of not treating obesity as a single clinical problem. If we treat BED patients only with weight-loss interventions, we won't target the actual problem, as B's case suggests. While BED may occur in normal-weight individuals, the patient in such cases is typically in the early stages of the disorder. For early BED patients, prompt intervention may stop the binge-eating behavior before medical and psychological consequences become irreversibly severe.

Research also suggests that patients binge eat to distract themselves from uncomfortable feelings. Indeed, binge eating is often viewed as emotion-driven eating, done in response to anxiety, depression, and/or boredom.[4] This has certainly been the case for B, who reports difficulty connecting to others since childhood as well as identifying and allowing herself to experience emotions.

Lessons from Neurobiology

What might happen in the brain to drive this kind of extreme eating behavior? Even in healthy individuals, eating is a complex process, involving not just physiological but also psychological and emotional processes. We eat not just because we need sustenance but also because food tastes good, we're with friends who are also eating, and/or we're bored and food is available.

Three neural pathways interact to drive eating behavior: One codes for the perceived salience, or importance, of a food stimulus; another codes for the rewarding sensation of actual eating; and the third helps us control our consumption based on considerations of both short- and long-term outcomes, such as weight gain.

Basic sensory information about food is processed by a brain region called the insula, along with the frontal operculum. The insula also plays a role in networks that determine salience. To identify and evaluate the rewarding properties of the food, closely connected brain regions collectively known as the ventral striatum, the putamen, and caudate, are called into action. Finally, the circuitry that helps us control our responses to food includes the dorsal caudate and dorsal anterior cingulate cortex, ventrolateral prefontal cortex, and parietal cortex.[5] Together, these pathways combine, in effect, to weigh the salience of a stimulus, its reward value, and the longer-term consequences of consuming it, and thereby determine eating behavior.

What's different about these pathways in BED? Generally, overeating is attributed to an elevated experience of reward, a reduced ability to inhibit the drive to eat, or some combination of the two. Typically, studies of these pathways in BED patients compare the latter to weight-matched controls as well as lean participants. While some of the research is conflicting, overall we see decreased inhibitory control and increased sensitivity to reward.

If we start to parse out different types of self-control, BED is associated with a range of deficits. Individuals with BED tend to do worse on tasks related to motor inhibition and attention, and these deficits seem to be related to binge-eating severity rather than obesity.[6] While it's unclear whether increasingly severe binge eating leads to inhibitory control problems or vice versa, these impairments certainly help maintain the disordered eating. If individuals are more susceptible to impulsive decision making, their vulnerability to binge eating increases, and breaking the cycle of binge eating becomes more difficult.

However, we can't just tell BED patients that they need to make better choices and exert more willpower, given the differences in the neurobiology that underlie inhibition. Compared to both lean and weight-matched control participants, BED patients appear to have abnormally low activity in impulse-control related frontal brain regions. A recent neuroimaging study asked BED participants to complete a task in which they had to resist reading a word in favor of naming the color of the word's printed type—a classic test of inhibitory control.[7] The study showed reduced activation in the ventrolateral PFC, inferior frontal gyrus, and the insula. Further, the more impairment that BED participants had in recruiting those neural pathways involved in self-control, the worse they were at dietary restraint.

Studies of the relationship of reward sensitivity to BED are more numerous, and typically point to an elevation in the brain response to both food and non-food reinforcement. Functional magnetic resonance imaging (fMRI) research has shown links between aberrant brain activation and behavioral tasks of delaying gratification as well as self-report measures of sensitivity to reward. Some fMRI studies have lumped BED patients with non-bingeing obese subjects and, unsurprisingly, have reported inconsistent findings. Overall, findings have tended to indicate increased reward sensitivity, particularly just after having eaten. Those studies that have looked at more pure samples of BED patients have found exaggerated responses in the orbitofrontal cortex to food reward compared to obese or lean controls.[8]

If we distinguish between viewing images of rewarding things and anticipating or receiving actual rewarding things, the findings again tend to vary. During the anticipation of monetary rewards, for example, BED subjects show less activity in the ventral striatum compared to obese non-bingeing subjects—though they show no significant difference on this measure from lean control subjects.[9]

Wang and colleagues[10] have demonstrated, using positron electron tomography (PET), that seeing, smelling, and tasting a food (but not eating it) significantly increases dopamine in the caudate in BED participants. Dopamine is a neurotransmitter that helps mediate the motivation to eat and, in the dorsal striatum, is associated with habit learning in which behavior becomes automatic and no longer necessarily linked to pleasurable outcomes. The Wang study linked higher caudate dopamine levels to binge -eating severity rather than to weight, supporting the idea that separate brain circuits underlie eating pathology and the development of obesity.

Considering the two prior findings, in the context of the relevant literature, we could speculate that participants do not fully anticipate the effect that eating will have on their psyche, eating only out of habit. Thus, they become overwhelmed upon receiving the food, potentially driving the loss of control that seems central to binge eating. While the research in BED is still limited, similar results suggesting a blunted anticipatory response to reward has also been seen in substance abuse.

So how do these neurobiological differences manifest themselves in personality? Do characteristic temperament traits exist prior to the development of binge-eating behavior? Is there any way to predict the development of the disorder? A recent meta-analysis suggests that BED patients may demonstrate slightly elevated harm avoidance (a measure of anxiety, inhibition, and inflexibility) and potentially novelty seeking (a facet of impulsivity).[11] Other studies have shown that impulsivity—specifically negative urgency, or the tendency to act impulsively when distressed—may be related to the emergence of binge eating.[12] This seems consistent with the emotional dysregulation that is typically reported in this population—dysregulation that binge eating may be aimed at correcting. Overall, these two findings suggest that some level of impulsivity or harm avoidance may predispose some people to binge eat. These temperament traits also have been linked to anorexia nervosa and bulimia nervosa.

Binge-Eating Treatment

There are a number of empirically based treatment approaches for BED that share the goal of reducing binge-eating episodes and the accompanying psychopathology. The gold standard is cognitive-behavioral therapy (CBT), which targets maladaptive eating behaviors through a combination of self-monitoring, normalizing eating patterns, and building behavioral and cognitive coping skills.

Cognitive strategies involve identifying and restructuring cognitive distortions that help maintain binge-eating behaviors. For instance, the thought that "I've already broken my diet with this slice of pizza, so I might as well have the whole pie" is an example of "all or nothing thinking," and can be reframed as, "I may have had a slice of pizza, but I can still eat healthily the rest of the day." CBT also targets maintenance of normal eating behavior and relapse prevention.

However, some research indicates that only 50 percent of BED patients who complete CBT treatment experience lasting improvement, and an estimated 33.3 percent of patients continue to meet criteria for eating disorders five years after beginning therapy.[13] These are dismal numbers, but there is evidence that a longer course of treatment may show better results.[14]

Given that binge eating is widely seen as a means of managing unwanted emotions, some treatments have targeted patients' emotional regulation abilities. These treatments also have shown promise. For example, dialectic-behavior therapy (DBT) is a cognitive-behavioral treatment approach, originally developed for borderline personality disorder, that focuses on mindful awareness and tolerance of emotion, building coping skills, and interpersonal effectiveness. In studies of DBT's effectiveness, most participants significantly decreased their binge-eating behavior.[15] While differences compared to a control group that received nonspecific treatments were not significant after follow-up, an additional study of DBT-based guided self-help suggests that persistence of binge-eating abstinence is influenced by changes in emotional regulation.[16]

Pharmacological treatments that target brain chemistry directly have shown mixed success. A meta-analysis of pharmacotherapy for BED suggests a medium effect of antidepressant use for the reduction of binge-eating behavior.[17] Studies suggest that combining pharmacotherapy with CBT or interpersonal therapy (IPT) does result in a significantly improved response rate, but combination treatments may not be any better than psychotherapy alone. In general, for prospective BED drug therapies there has been a relative lack of controlled clinical trials in large samples of patients.

Lisdexamfetamine (Vyvanse), an appetite suppressing amphetamine, is currently the only medication that is FDA-approved specifically for treating BED. Like other stimulants, lisdexamfetamine works by increasing extracellular dopamine and norepinephrine, and while the mechanism of action is not definitively known, it is possible that increasing synaptic dopamine in the striatum may weaken or reduce the characteristic under-response to anticipation of palatable food seen in BED. Clinical trial results indicate that it leads to the cessation of binge-eating behavior in up to 50 percent of treated subjects.[18] However, there are significant side effects, experienced by over 80 percent of participants on the medication, and there is a significant risk of abuse and dependence. Furthermore, it's unclear that lisdexamfetamine works for people who have depression or other medical issues related to obesity—as most BED patients do. Psychological and behavioral treatments are necessary to fully address the complex psychological and medical syndrome associated with BED.

If pharmacotherapy is not a magic bullet, how can we improve cognitive behavioral treatment to target the neurobiology? Reducing the rewarding properties of food is one possible approach. It may be difficult, but preliminary research in obesity suggests the possibility of habituating patients to food through repeated exposure without actual consumption, thereby reducing cravings.[19] The same research suggests that control over eating behavior may be strengthened by a combination of stimulus control—such as minimizing exposure to binge foods, changing environmental cues for binge eating, and reducing vulnerability factors for binge eating, such as extended periods of time alone—and inhibitory training. Both approaches may help individuals develop the needed skills to resist urges to initiate binges or stop them once they've started.

We might also consider working with the temperament traits that promote binge eating rather than against them. If individuals with BED are more reward-sensitive, then frequent and salient rewards for abstinence during therapy may be more effective than punishments for lapses. Rewarding social experiences that don't involve food may also reduce reliance on food, as well as potentially reducing any underlying depression.

We propose that further advances in BED treatment can come from behavioral interventions that target BED's apparent neurobiological mechanisms. Anecdotally, this strategy appears to be working for B: She has spent four months in a partial hospitalization program using DBT and CBT treatment to advance her learning skills and become more mindful of her emotions, as well as better tolerate distress and improve interpersonal relationships. While she hasn't lost any weight, she has binged only a handful of times. Meanwhile, she's learning to cook for herself and eat regularly, rather than depriving herself or following fad diets. She's also worked significantly on regulating her emotions: "Learning other ways of dealing with emotions improved my life one-hundred percent. If I lose weight because of it, that is a bonus. If not, at least I have my life back."

11

Failure to Replicate

Sound the Alarm

By John P.A. Ioannidis, M.D., D.Sc.

John P.A. Ioannidis, M.D., D.Sc., holds the C.F. Rehnborg Chair in Disease Prevention at Stanford University, is professor of medicine, and of health research and policy, and is director of the Stanford Prevention Research Center at the School of Medicine. He is professor of statistics (by courtesy) at the School of Humanities and Sciences, one of the two directors of the Meta-Research Innovation Center, and director of the Ph.D. program in epidemiology and clinical research. Ioannidis, who grew up in Athens, Greece, received a doctorate in biopathology from the University of Athens and trained at Harvard and Tufts (internal medicine and infectious diseases), then held positions at the National Institutes of Health, Johns Hopkins, and Tufts. He has served as president of the Society for Research Synthesis Methodology.

Editor's Note: Science has always relied on reproducibility to build confidence in experimental results. Now, the most comprehensive investigation ever done about the rate and predictors of reproducibility in social and cognitive sciences has found that regardless of the analytic method or criteria used, fewer than half of the original findings were successfully replicated. While a failure to reproduce does not necessarily mean the original report was incorrect, the results suggest that more rigorous methods are long overdue.

PSYCHOLOGICAL SCIENCE HAS BEEN a highly prolific discipline. Compared with other scientific fields, it has had one of the highest rates of experimental "success." Analyses have shown that almost all studies in the field (90 to 100 percent) claim statistically significant results with p-values (which indicate the likelihood that the experiment's outcome represents mere statistical "noise") of less than 0.05.[1,2]

This may sound like a cause for celebration: Success seems to be ubiquitous! In fact, it should be a cause for concern. Other analyses have shown that the statistical power of studies in the field is too modest on average[3-5] to account for such a high success rate. In other words, the statistical "noise" inherent in these studies has been so high that it should have caused many more negative results than were reported—even if all the hypotheses targeted in these studies were true.

The lack of replication is even more worrisome. Psychological science has one of the lowest rates of replication studies, in particular exact replications by independent investigators. A recent text-mining survey of the 100 most-cited psychology journals since 1900 found that only 1.07 percent of the published papers were categorized as replications.[6] Some replications may have been missed by the survey authors' automated search, but almost certainly not many. Of the identified replications, only 18 percent were true replications, the remainder being extensions of the work using different methods, settings, populations, or other deviations ("conceptual replication").[6] Furthermore, only 47 percent of the identified replications were done by investigators who were not authors of the original studies.[6]

Such figures reflect psychological science's incentive structure, in which replication experiments have been relatively unwelcome: They have attracted little funding, typically have been harder to get published, and have received little academic recognition.

Several scientists have argued that this is a recipe for disaster.[7-9] Indeed; this author has proposed that in scientific fields where underpowered experiments are the norm and significance-chasing behavior thrives, one would expect the majority of "statistically significant" results to be false-positives.[7]

Conceptual replication can offer complementary insights, but cannot replace direct, exact replication. When there are many unsorted false-positives, conceptual replication with a pressure to find more significant results may simply perpetuate fallacies and lead even more investigators astray.

Replication done by the original authors may also have value, but can lead to a culture of "inbreeding"[10] in which each scientific finding is reproducible only within a restricted environment—the laboratory of a professor and his or her team and mentees. Outsiders who attempt to enter this closed world may "spoil" the results, much like explorers who have entered ancient sealed tombs only to find that beautiful colored frescoes within are blanched by the contact with fresh air.

A Game Changer?

Theoretical concerns and sporadic evidence have not been able to convince the field to change its dislike for exact replication. However, this pattern may now change, because far more powerful, and hopefully more convincing, evidence has emerged from the Reproducibility Project, led by the Center for Open Science in Charlottesville, Virginia.

In this four-year project,[11] 270 experienced investigators joined forces to conduct exact and adequately powered replications of 100 studies that had been published in three leading psychology journals. The exercise was carried out with exemplary rigor and involved close communication with the original authors to ensure that the replication adhered as faithfully as possible to the original experimental conditions. There are different statistical approaches to define successful replication, but all of these suggested that nearly two-thirds of the original findings were false-positives, with worse performance in social psychology than in cognitive psychology.[11]

For example, one replication study tried to replicate whether participants primed with close spatial distances would report stronger feelings of closeness to their family, siblings, and hometown than participants primed with long distances, as proposed in an earlier paper published in 2008.[12] Despite using identical stimulus materials, dependent variables, and analysis strategies, the replication effort could not replicate the original findings on spatial priming and emotional closeness. Another replication study aimed to replicate that reduced self-regulation resources correlate with increased biases in confirmatory information processing, as previously published.[13] The original paper had shown that the depletion of self-regulation resources influences the search and the processing of standpoint—consistent information in a personnel decision case, even when confronted with an alternative explanation—i.e., the ego threat, with associated failure cognitions and negative emotions. This could not, however, be documented in the replication effort. More examples and details on the studies replicated can be found in https://osf.io/ezcuj/.

The results of the Reproducibility Project caused a flurry of interest from the scientific community and the general public.

The Reaction

Some of the immediate responses were wrong or counter productive. On one extreme, commenters suggested that psychology is not a science and should be abandoned or be called an art. On the other extreme, some dismissed the failures to replicate as having been due presumably to unknown differences between the original experimental setups and the replication attempts.

Let's not spend time arguing whether psychology is a science. It is a very important science, and, as the Reproducibility Project reminds us, has been at the forefront of the study of the scientific method and its biases.

Lack of replication and reproducibility has been documented in other scientific disciplines.[14-16] In fact, those that have recently started performing replication experiments have seen very high replication failure rates, even higher than those of the Reproducibility Project in psychology.[17,18] Meanwhile, disciplines that have adopted replication in large scale—e.g., genetic epidemiology—have seen dramatic improvements in the reliability of their results.[19] Many fields of neuroscience and neurobiology are characterized by the conduct of small, underpowered studies,[5] and reproducibility is likely to be low. Thus, what we have documented in the Reproducibility Project may be a pattern that affects many other disciplines.

It is also easy to refute the suggestion that unknown experimental differences are the chief cause of irreproducibility. If that were the case, one would have seen larger as well as smaller effects in the replication studies-compared to the originals. In fact, the replication effect sizes were almost always markedly lower in the replication efforts, rendering them statistically non-significant.

Probably most of the replication failures in psychological science are due to bias in the original results. It is not possible to pinpoint exactly which specific study was biased and how bias exactly happened—replications may also have been biased occasionally. However, the notion that all results are correct despite failures to reproduce them amounts to irresponsible hand-waving. If we want a research finding to make any claim to generalizability or, better yet, to be used for practical purposes, other scientists should be able to reproduce it relatively easily. No one would like to fly in a plane that has flown successfully only once, especially if its manufacturers are satisfied that it flew only once and don't mind that it may crash on its second flight. And of what use would a plane be if it flew once and was dismantled, and afterward no one could rebuild it?

In the Reproducibility Project, one-third of the 147 studies that were identified as possible targets for replication were not picked by any of the 270 replicators, since they were felt to be too difficult, if not impossible, to even try to replicate.[11] It is unclear what the value of research is when no one, other than the original scientists, can ever approach it. Among the studies in the project for which replications were made eventually, difficulty in building the replication experiments was a predictor of replication failure.[11]

The Reproducibility Project will hopefully lead to a better appreciation of the need for incorporating exact replication more routinely in the life cycle of research in psychological science. There is clearly a need for more replication studies, done by independent investigators. There are, however, still many unanswered questions and concerns about how to optimally implement a replication agenda.

Other Considerations

One major concern is the level of resources required. Doing replication well takes a lot of effort. Hastily conceived, suboptimal efforts may even do harm by generating spurious results and confusion. A replication agenda will require substantial funding. While this may be seen as eroding a discovery budget that is already constrained, such a perspective would be misleading. Replication is not some sort of unfortunately imposed policing; it is actually an integral part, perhaps the most integral part, of the scientific discovery process. If the current situation is such that the majority of "discoveries" are false, then replication is the most essential element in any true discovery. Replications also allow us to identify rapidly the avenues of research that warrant further investigation and have the best potential for future yield. In short, more reliance on replication can help save us from fund-wasting dead- ends and false-positives.

Should everything be subjected to replication? Some other scientific fields have accepted this as a norm. In genetic epidemiology, for example, it is impossible to publish anything in a high-profile journal without independent replication. However, in the field of psychology there may be insurmountable barriers to the adoption of this principle. These include practical difficulties (e.g., for very complex experiments). Also, the community may not be ready for such a sweeping paradigm shift. It may be necessary to target replication efforts in a more limited, strategic way.

As a first step, research could be categorized as "replicated" or "unchallenged."[20] Unchallenged research would have to be treated with extra caution—as more likely false than true, perhaps with substantial variability across subdisciplines. Samples of studies of different types, and stemming from different subdisciplines, would be subjected to replication periodically to examine what is the current replication performance of the subdiscipline. Therefore, one would know that working in field X with study design Y carries a Z percent risk of non-replication. Such figures would change over time, particularly if the field were to adopt more safeguards to improve its overall research practices. These safeguards could include registration of research protocols prior to experiment, data sharing, team science approaches, or other practices that improve transparency, efficiency, and reliability.[21-23]

For the more influential and heavily cited studies, the imperative for independent exact replication should be very hard to resist; these studies should be subjected to replication. It would make little sense to neglect to replicate a study upon which hundreds or thousands of other investigations depend.

Finally, studies that aim to inform practical applications or otherwise affect humans, such as treatments for psychological problems, should have thorough replication as a sine qua non before being adopted in everyday practice.

At a first stage, such a replication science agenda is also likely to require a very small amount of funds, perhaps 3 to 5 percent of the current research budget—a bargain if it reduces the 50 to 90 percent of the research budget that currently seems to be wasted on irreproducible research. That said, the devil can be in the details, such as who will fund replications, when they-should take place, and how they should be conducted. Editors and reviewers also need to become friendly to good replications[24,25]; publishing replications will only greatly encourage replication.

To get where we need to go, all action plans will need to have strong grassroots endorsement by the scientific community. The Reproducibility Project, and the favorable responses to it, show that many scientists care deeply about making research more reproducible. There is no reason to doubt that the general public would also want the same.

12

Cognitive Skills and the Aging Brain

What to Expect

By Diane B. Howieson, Ph.D.

Diane B. Howieson, Ph.D., is a neuropsychologist and associate professor emerita of neurology at the Oregon Health & Science University. She was head of the neuropsychology division of the Portland VA Health Care System for 12 years before joining the C. Rex & Ruth H. Layton Aging & Alzheimer Disease Center at the Oregon Health & Science University. She has published more than 40 scientific articles, primarily in the areas of aging and dementia. Howieson is co-author of the widely used professional book: *Neuropsychological Assessment.*

Editor's Note: Whether it's a special episode on the PBS series, The Secret Life of the Brain *or an entire issue dedicated to the topic in the journal* Science, *a better understanding of the aging brain is viewed as a key to an improved quality of life in a world where people live longer. Despite dementia and other neurobiological disorders that are associated with aging, improved imaging has revealed that even into our 70s, our brains continue producing new neurons. Our author writes about how mental health functions react to the normal aging process, including why an aging brain may even form the basis for wisdom.*

EVERY DAY WE PERFORM hundreds of cognitive tasks but are mostly unaware of the effort involved. These tasks take different forms, such as noticing colors, remembering names, or calculating time on a watch. Measures of brain function using functional magnetic resonance imaging (fMRI) show that the most active areas of the brain vary according to the task being performed. The data confirm what researchers have known for many years: that our mental functions are composed of many distinct types of cognitive abilities.

Mental abilities change throughout life, first as a result of brain maturation and later with aging of brain cells and their billions of complex interconnections. As people age, their movements and reflexes slow and their hearing and vision weaken. Until the 1990s, most aging research examined cognitive abilities of adults younger than 80. More recent research includes the fast-growing 80s-and-older population and has advanced our understanding of cognitive changes in the elderly. Scientists in a recent study asked, "When does cognitive functioning peak?"[1] and found evidence for considerable variability in the age at which cognitive abilities peak and decline throughout life.

Cognitive Changes with Aging

Certain cognitive abilities show at least a small decline with advanced age in many, but not all, healthy individuals. Although differences between the young and elderly can be shown in some cognitive areas described below, declining ability does not translate into impairment of daily activities. These changes are subtle. The most consistent change is cognitive slowing. For example, on a writing task in which people were asked to substitute as quickly as possible symbols for numbers, 20-year-olds performed the task almost 75 percent faster on average than 75-year olds. Age-related slowing is also evident on certain attentional tasks, such as trying to grasp a telephone number when someone rattles it off quickly. Overall, cognitive slowing is thought to be a contributing factor in elderly people's higher rate of automobile accidents per miles driven.[2]

Age hinders attention, particularly when it is necessary to multitask. When switching from one task to another, the elderly have more difficulty paying attention to multiple lanes of traffic, for example, or noticing if someone is about to step off a curb at a busy intersection. Processing information rapidly and dividing attention effectively are cognitive skills that peak in young adulthood. How fortunate it is that college and vocational students are typically at an age when the brain is working with optimum efficiency.

Similarly, the ability to keep multiple pieces of information in mind at the same time is another skill that peaks around ages 18 to 20 and becomes more difficult thereafter. Every time you mentally calculate a tip in a restaurant, you use an information processing skill called "working memory." In the clinic, we often test working memory by asking people to recite backward a string of numbers that we have just read to them. This task requires working memory because the numbers need to be held in mind long enough to rearrange them.

While memory declines for many people over time, the exact nature of the decline depends on the particular type of memory. To be able to recall an event or new information, the brain must register the information, store it, and then retrieve it when needed. The ability to recall new information, such as reading material, peaks early and gradually becomes more challenging after age 40, particularly for visual material. Studies show that by age 70 the amount of information recalled 30 minutes after hearing a story once is about 75 percent of the amount remembered by an 18-year old.[3] Recognizing information from the story is easier than remembering it without any cues, and this ability is usually well retained throughout life. In other words, older adults are less likely than young adults to freely recall most of the information from a recently read news article, but they may be just as good at recognizing the content if someone talks about it.

Language skills develop rapidly throughout childhood and are well retained throughout adulthood, with one exception: Recalling a familiar person's name or a particular word during conversation commonly becomes harder for adults after age 70. Although this type of word-finding involves memory, the problem lies in accessing the word, even though it exists in the person's knowledge or vocabulary. Sometimes this problem manifests as a "tip of the tongue" experience—you feel close to recalling the word and may even know that it begins with a given letter. If someone else says the name or word, you easily recognize it. This problem is particularly frustrating because mustering effort to force recollection of the word is likely not to help. The word is apt to pop into mind later when you are going about another activity.

Visual perceptual abilities, principally the ability to understand spatial relationships, also show decline with age, especially after age 80. This weakness causes another driving-related problem, such as not knowing how far away a curb is or how much to turn to parallel-park a car. Visual scanning ability also can diminish so that, for example, it becomes more difficult to see a misplaced object among other items.

Executive functioning refers to higher-level skills, such as conceptualizing a problem, making appropriate decisions, and planning and carrying out effective actions. Older adults tend to be slower in conceptualizing problems and less ready to change strategies when circumstances shift. In one well-known study involving decision making, approximately one third of older adults did poorly compared to younger adults.[4]

The task used to assess decision making in this study was the Iowa Gambling Task. Each participant is asked to choose cards one at a time from one of four decks. Some cards bring rewards in the form of play money, and others bring penalties. The goal is to win as much money as possible. Deck selection is based on hunches, and subjects are given immediate feedback. The catch is that the decks have different mixes of reward and penalty cards. In two decks, the reward cards bring relatively low rewards, but the penalty cards bring relatively low penalties, and choosing from these decks brings a net gain in the long term. In the other two decks, the reward cards bring higher rewards, while the penalty cards bring higher penalties too, so that choosing from this deck brings a net loss long-term.

Over time, in the study, young participants tended to shift their performance so that they no longer selected cards from all decks—they selected only from the decks that gave them a net gain. However, a subset of older participants continued to select cards from all decks, resulting in a net loss. These same older participants also were more fooled by deceptive advertising in another experiment.

Findings from studies like these give a possible explanation of why some older adults are at greater risk of falling prey to fraud. Yet for many older adults, difficulty thinking through problems and mental flexibility may not be noticeable until the 80s or beyond.

Cognitive Optimism

Other important cognitive abilities decline little if any with age. Language and vocabulary are well retained throughout the lifespan. In fact, vocabulary continues to improve into middle age. Recall of general knowledge acquired at a young age and well-practiced skills like arithmetic also peak in middle age and are resistant to age-related decline. In general, these age-resistant cognitive skills have been strengthened by experience, including situations that require reasoning and judgment. For example, if asked why many foods need to be cooked, most adults will have no trouble answering based on a lifetime of experience. In addition, older adults often have a better overview of a situation, or better appreciation involving the impact of a single event, than younger people do, because of their greater life experience.

The cognitive operations described above do not exist in isolation. Multiple cognitive skills, such as attention, memory, and reasoning, are involved in performing even simple daily tasks. Some activities require a complex combination of cognitive skills. Among these activities are the social behaviors of everyday life used while shopping, riding a bus or train, dealing with neighbors, or helping a friend. In fact, social skills strongly depend on the cognitive ability to form accurate impressions of others.

Although we have a good understanding of most of the cognitive changes that tend to occur with aging, we understand relatively little about age-related changes in the "social cognition" that we use during social interactions. Social behavior relies on a combination of cognitive and emotional factors, and the influence of aging on these factors is multifaceted. For example, a social impression—an impression of a person one has just met—is built up from factors such as physical appearance, voice quality, facial expressions, and ways the person is behaving. Even though older adults have more limited information-processing capacity, their automatic perceptions of people seem to be intact.

Most of us, particularly as we get older, have had the experience of meeting someone new and a minute later not being able to recall their name. Although usually assumed to be a memory failure, this is actually a failure to fully attend to the name because one is distracted by the broader social interaction, so that the name is not strongly registered in memory. (The trick here is to repeat the name aloud as soon as you hear it, as a confirmation, and then repeat it to yourself silently a couple of more times within a few minutes.)

In this situation, an age-related decline in information-processing speed tends to handicap the older adult. Similarly, age limitations interfere with performance when information acquired in an unfamiliar situation needs to be processed quickly or there are distractions that should be ignored. As a result, older adults on average consider fewer bits of information and use less effective decision-making strategies when they are in unfamiliar situations compared to younger adults.

One view supported in studies of social cognition is that older adults' limitations in the amount of information that can be processed quickly and accurately are often counterbalanced by increased social expertise from accumulated experience and knowledge.[5] Thus, in familiar situations, middle age and older adults, compared to younger adults, tend to make more accurate interpretations of the behaviors of others when prior experience and knowledge helps to focus attention and make it more efficient. A lifetime's worth of experiences in social situations can facilitate decision making—and that is sometimes referred to as having wisdom.

Individual Differences in Aging

For most people, cognitive or social decline with aging is minor and influenced by multiple factors. One factor is called "cognitive reserve."[6] People who are more intelligent at a young age or have better cognitive maintenance through education, occupation, or stimulating activities retain cognitive skills with aging better than those who are less accomplished in these respects. A recent study involving a large number of people in Scotland who had intelligence tests at age 11 and again 50 years later found that the biggest predictor of cognitive ability at the older age was cognitive ability at age 11.[7] It is possible that being blessed with the right genes accounted for much of this benefit, although little is known about what genes might be involved.

Having friends and enjoying activities with others also appears to be beneficial. Numerous studies have shown that the level of social engagement, such as the size of a person's social network or frequency of contacts, promotes cognitive health or reduces risk of dementia.[8] Having a purpose in life has been shown to be associated with reduced risk of Alzheimer's disease. Together these factors help explain the variability we see in how well cognitive function is retained with advanced age.

Normal Brain Aging

Years ago it was widely assumed that the death of neurons, cells that transmit signals throughout the brain, is a common part of aging. We now know that there is little evidence of this. However, the brain does tend to get smaller as people age and a number of changes appear to account for this decrease in size. Each neuron has a cell body and a number of processes called dendrites that extend in many directions toward other neurons for receiving signals. Think of a tree limb with many branches. During aging the size and complexity and efficiency of this "arborization" decreases, making communication between cells less effective. Each neuron also has an axon that transmits signals from one cell to another; these axons make up the "white matter" in the brain. Damage to white matter tracts due to aging contributes to decreased brain size. These and other structural brain changes associated with aging correspond to age-related differences in performance across cognitive tasks. For example, white matter deterioration in the front of the brain has been associated with slower information processing speed and more difficulty recalling information.

A number of other structural and chemical changes in the brain that occur with aging are not fully understood. One condition is the buildup of a small neuronal protein fragment called amyloid beta, which accumulates to form aggregates of various sizes. Dense conglomerations or "plaques" of these aggregates are characteristic of the brains of Alzheimer patients and are seen to a lesser extent in elderly people with milder cognitive impairment. Alzheimer patients also develop an abnormal version of the neuronal protein tau within brain cells. Converging evidence from multiple types of studies suggests that amyloid plaques trigger the buildup of abnormal tau, which causes neuronal loss that is associated with dementia. Because some adults with dense amyloid beta plaques appear to have normal cognitive function, the relationship between the two is not fully understood. These adults may have an unknown protective factor. More information is needed from longitudinal studies to know whether the accumulation of amyloid beta aggregates in cognitively intact adults will lead to declining cognitive ability and ultimately Alzheimer's disease if a person lives long enough.

With aging come increased risks for vascular disease for many people. High blood pressure, high levels of good cholesterol (HDL), high triglyceride levels (a type of fat found in the blood), obesity, and diabetes increase risk of stroke and white matter disease. Keeping the brain healthy through good nutrition and physical activity is important to reduce the risk of cognitive decline associated with vascular disease. A healthy diet includes limiting the intake of sugar and saturated fats, particularly trans fats. Scientists are learning the many ways in which physical exercise affects the brain, ranging from benefits shown in animal models at the cellular level, such as stimulation of brain-growth factors or reduction in oxidative stress, to a decrease in white matter damage in humans.

An exciting newer area of research, made possible by technical advances in imaging, is the study of age-related changes in brain activity. It is now possible, for example, to monitor brain activity by measuring how much oxygen (fMRI scanning) or sugar (PET scanning) individual brain regions consume. Even simple acts cause widespread activation of multiple brain regions and, by studying the pattern of active areas during a cognitive task, researchers can learn which networks or regions are task-specific. During normal aging, changes occur in the pattern of stimulation of neural networks, causing increased activation in some areas and decreased activation in others. Studies reveal that when an elderly person performs a cognitive task at the same level as that of a young adult, more areas of the former's frontal brain regions "light up," suggesting more brain activity is needed to maintain cognitive performance. Many questions still need to be answered in order for science to understand the full impact of aging on brain network function.

Overestimating the Impact of Aging?

It is unclear how much cognitive decline is purely the result of aging of an otherwise healthy brain. Older adults in cognitive studies are more likely than young or middle aged adults to include people with undetected Alzheimer's dementia, cerebral vascular dementia, and other brain diseases that are more prevalent in people over 70. In one recent long-term study of older adults, brain diseases detected at autopsy accounted for a large amount of the cognitive decline measured during the study.[9] If individuals with undetected early-stage brain diseases are included in studies of normal aging, the amount of measured cognitive decline that is purely age-related will be exaggerated.

Around the world, researchers have made significant progress in understanding factors that influence cognitive health and the related risk of developing dementia. This research has been given high priority because of the devastating personal, family, and societal costs of illnesses such as Alzheimer's disease. Advances have led to an understanding of some genetic and environmental factors but much is still unknown. Currently, an intensive quest is under way to find new treatments to stop, slow, or even prevent Alzheimer's and related disorders that cause dementia. Providing researchers with the funds necessary to make progress in this search will hopefully lead to the discovery of better ways to reduce late-life cognitive decline.

BOOK REVIEWS

13

Review: Neal Barnard's *Power Foods for the Brain*

By David O. Kennedy, Ph.D.

David O. Kennedy, Ph.D., is professor of biological psychology and the director of the Brain Performance and Nutrition Research Centre at Northumbria University in the United Kingdom (www.nutrition-neuroscience.org). His research involves the effects of nutritional interventions, including vitamins and minerals, omega-3 fatty acids, amino acids, and a host of plant-derived extracts and compounds, including polyphenols and caffeine, on human brain function. His recently published book, *Plants and the Human Brain* (Oxford University Press), describes the effects of a multitude of plant-derived compounds on brain function.

Editor's Note: Can a plant-based diet help stave off dementia and Alzheimer's disease? Neal Barnard, M.D., president of the Physicians Committee for Responsible Medicine who was featured in the popular documentary Forks over Knives, *makes a case for it in his best-selling book,* Power Foods for the Brain *(Hachette). In his review, David O. Kennedy leans on his own research on plants, nutrition and human brain function to discuss the merits of the author's claims.*

THE COMPLEX INTERACTION BETWEEN numerous components of our diets and the functioning of our brains is a fascinating topic. The physiological and biomolecular effects of consuming a range of nutrients, including but not restricted to vitamins, minerals, omega-3 fatty acids, amino acids, and numerous polyphenols and other phytochemicals, are receiving unprecedented attention by researchers. Equally, the detrimental effect of the typical contemporary high-energy and low-nutrient diet is becoming better understood.

The emerging pattern of findings suggests that potential benefits of consuming specific nutrients and adopting a healthier diet can range from almost immediate boosts in cardiovascular and brain function to neuroprotective effects that may be able to extend good brain function deeper into very old age. If you are interested in arriving at a better understanding of this wide, complex, and fascinating topic, then *Power Foods for the Brain* is probably not really the book for you. If, on the other hand, you fall into the book's apparent intended demographic—aging, nonexercising, possibly overweight, or prediabetic with a poor diet, who worry that their mental faculties are deserting them—then this book may well be what you are looking for.

Neal Barnard's self-help book certainly seems aimed firmly at anybody who checks a few of those boxes, and for this group it certainly provides a practical guide to a lifestyle change that almost certainly will pay dividends, in terms of both general health and long-term brain function. In essence, Barnard argues for a strict vegan diet composed entirely of plant-derived foods, a bit of physical exercise, avoiding potentially toxic metals, and engaging in mental exercise. The first two of these will undoubtedly prove beneficial, and the latter two, while unsupported by any clear evidence at present, should do no harm beyond the costs of purchasing new cooking pots and brain-training computer software.

My only reservation regarding the book's intentions involves research showing that, for public-health messages, anxiety is only an effective driver of change if the proposed remedy is achievable by the audience. If it isn't, individuals will engage in an entirely internal process of anxiety-reducing psychological self-defense, typified by denial that the fear-inducing outcomes apply to them at all. The message in *Power Foods for the Brain* is delivered by increasing the reader's anxiety about the prospect of their imminent descent into cognitive dysfunction and dementia. However, I wonder whether if Barnard's all-or-nothing, hard core vegan remedy, which disallows all meat, fish, and dairy products, may simply be seen by the typical intended reader as being unachievable. It may be difficult to follow for anyone who feeds a family or partner, and it will inevitably involve a major life-changing modification of eating habits, including avoiding the processed, ready-made foods in the typical diet in favor of cooking and preparing all foods from healthy, plant-derived ingredients. The diet may be well worth following, but if the remedy being proposed doesn't seem achievable (and for many it might not), it isn't likely to be adopted.

In reality, it isn't necessary to become a vegan to benefit from improving your diet. Simply shifting your dietary habits so that you avoid the bad aspects of our contemporary diets—energy-dense foods, "bad" fats, and refined sugars—while increasing your consumption of fruits, vegetables, and other plant-derived foods, will pay dividends in terms of potential improvements to cardiovascular and metabolic health and improved or preserved brain function, particularly if coupled with increased exercise. So, maybe proposing a slightly softer remedy—one that acknowledges simply that the farther you travel down the road toward replacing the unhealthy elements of your diet with healthy plant-derived foods, the better—might be a more effective message at the end of the day.

Throughout, Barnard seasons his advice with a smattering of relevant science, and the scientific rationale for his dietary recommendations is written in an accessible, interesting style. He certainly has a knack for describing complex biology in understandable language. But where the science content falls down is in the very limited scope of the information; the book provides only a brief scientific rationale for reducing the consumption of "bad" saturated fats and trans fats and increasing consumption of "good" omega-3 fats. It also espouses building a "vitamin shield" by increasing consumption of foods high in vitamins E, B6, B9, and B12. The stated rationale for boosting the three B vitamins is their contribution to breaking down a naturally occurring, potentially neurotoxic amino acid called homocysteine, thus preventing its buildup and reducing its toxicity. However, it is unclear to date whether homocysteine is a causal factor in cognitive decline and dementia or merely a coincidental phenomenon related to less than optimal intake of these vitamins. Surprisingly, the book gives no consideration to the potential benefits of increased consumption of vitamins A, B^{1}, B^{2}, B^{3}, B5, B^{7}, C, or D, all of which have an equally strong claim in terms of potential benefits to brain function.

Even more surprisingly, there is no mention at all of the other bioactive phytochemicals richly expressed in the diet of plant-derived foods that the book espouses. For instance, the words "flavonoid" and "polyphenol" are not mentioned once, despite good evidence that these (and their other phenolic stablemates) are the classes of plant-food-derived compounds that may do the heavy lifting in terms of improved cardiovascular function and neuroprotection. Similarly absent are the obesity-inducing scourge of refined sugars and high-fructose corn syrup, and the complex interplay between brain function and the diet-induced deterioration in multiple cardiovascular and metabolic parameters. Simplification may be a necessary property of a self-help book aimed at a nonscientific audience—the message has to be understandable and clear, and not fogged by too much complexity—but while the proposed diet probably will work, it won't necessarily succeed for the stated reasons.

Another good argument Barnard doesn't touch is that your current brain function is, in part, a reflection of your lifelong diet to date, an argument that can be translated into clear societal advice to improve diets and the delivery of key nutrients from cradle to grave. While his book provides a potentially effective method of soothing some of the damage of a misspent life in dietary terms, it doesn't address this wider point at all. To me, this can only reduce the potential audience and therefore the impact of the book's message.

The recipes at the end of the book *do* look tasty. I didn't spend long hours in the kitchen trying them out, but I suspect my children might be on the receiving end of some of the healthy, sweeter options this summer (they'll definitely be trying the banana ice cream). But here, too, I would have liked to see more information on how to plan a sustainable vegan diet that contains adequate levels of all essential nutrients. All in all, the book is a useful guide to preserving brain function by modifying the diet (and getting that equally important exercise), but a little more of the fascinating science around this subject would have been appreciated.

14

Review: Leon N. Cooper's *Science and Human Experience: Values, Culture, and the Mind*

By Gary S. Lynch, Ph.D.

Gary S. Lynch, Ph.D., a professor of psychiatry & human behavior and anatomy & neurobiology at the UC, Irvine, has published over 600 papers, holds 25 patents, and co-founded two publicly traded companies. Lynch's work led the way to the modern theory of how synapses encode memory. This involves a change in the shape, and thus potency, of connections that is stabilized by a reorganization of the subsynaptic cytoskeleton. As might be imagined, the chemical pathways underlying these remarkably persistent synaptic modifications are complex. Lynch is also the co-inventor of "ampakines," a class of drugs that enhance memory and stimulate the production of growth factors. He is currently using ampakines in an attempt to reverse the negative effect of aging on the anatomy and physiology of brain cells. Lynch received his Ph.D. from Princeton University.

Editor's Note: Why are we reviewing a book written by someone who shared in the 1972 Nobel Prize in Physics for work on superconductivity? Because shortly after winning the prize, Leon N. Cooper transitioned into brain research—specifically, the biological basis of memory. He became director of the Brown University Institute for Brain and Neural Systems, whose interdisciplinary program allowed him to integrate research on the brain, physics, and even philosophy. His new book tackles a diverse spectrum of topics and questions, including: Does science have limits? Where does order come from? Can we understand consciousness?

Science and Human Experience is a collection of essays drawn from work published over a 40-year span by that remarkable theorist-scientist Leon Cooper. And these really are essays—reasonably short, often argumentative, filled with startling insights, and written from a personal perspective about issues of great interest. Cooper's writing style is charming, witty, and accessible; the essays, while filled with fundamental and complex questions, somehow wind up being fun to read.

Cooper is famous for his theoretical work on the phenomenon of superconductivity (roughly, an abrupt loss of electrical resistance at low temperatures), for which he was awarded the 1972 Nobel Prize in Physics, early in his career. Niels Bohr, Werner Heisenberg, and other giants had tried to develop explanations for the phenomenon, but resolution came only with the BCS theory of superconductivity that Cooper developed with his colleagues, John Bardeen and John Robert Schrieffer, who shared in the prize.

Later, Cooper became interested in what makes the brain a thinking machine. He doesn't really tell us why he took up this question, but we can find clues in the book's early essays on physics. Several essays deal with the nature of reality as knowable by humans, something that can't be addressed without some understanding of how individuals construct private versions of what's out there. I suspect that Cooper's walk on the wild side of science began with an itch to apply the awesome power of theory (with which he was so intimately familiar) to his own internal world. In any event, the essays in *Science and Human Experience* have a single, consistent viewpoint informed by both physical and life sciences. In this, the essays are unique.

The first of the book's three sections deals with interactions between science and society and, in particular, why the latter's attitudes about the former are often so negative. Cooper, rightly I think, argues that the public sometimes views the research enterprise as ultimately robbing society of its values and of leading to the replacement of belief and morality with cold empiricism. Of course, this issue takes its most contentious form in questions about the existence of God.

Cooper takes full advantage of the essay format, with its sanctioned inclusion of anecdotes, to lay out a clear position. A particularly telling story involves a meeting between the great mathematician Pierre-Simon Laplace and Napoleon Bonaparte. The emperor asks, "And where is God in your system?" and Laplace replies, "I did not need that hypothesis." Scientists, while happily postulating unobservables, steer clear of speculations that are not constrained by observables. The deity concept simply doesn't fit this criterion, but scientists, according to Cooper, never claimed that there is no God.

Problems also arise because people imagine science to be an activity far removed from everyday experience. Cooper argues instead that science emerges from a fundamental aspect of human nature: the need for explanation. Scientists satisfy this need by ordering things into a coherent picture, something that we all do when faced with complex, messy situations. What distinguishes science is that the structural relationships it discovers in the world are (largely) stable despite the evolving nature of the field; what changes is the explanation for those relationships. For example, Albert Einstein's theory for the elliptical orbits of planets supplants Isaac Newton's but doesn't dispute the trajectories described at the beginning of the scientific revolution.

Moving from the society to the individual, Cooper's second set of essays deals with the possibility of developing a physical explanation for thinking. Here we get a rare chance to watch the mind of a major theorist trying to resolve a very difficult problem. Starting with the idea that the brain consists of vast associative networks, Cooper's step-by-step analysis leads to an argument in which the brain 1) moves from the specific to the general in processing information and 2) inevitably leaps to conclusions. As such, it makes many mistakes but faces head-on the problem of insufficient data (generalization) and produces totally unexpected outcomes (leaps), operations that distinguish brain- versus computer-based thinking. He then pauses for what looks to be an epiphany: A critical function of education is to curb these potent, built-in tendencies in the machinery of thought (the book is filled with flashes of this kind). Put another way, as he attributes to Bertrand Russell, we are born as believers but learn to be skeptics.

These defining features of networks are by no means uniquely human but instead reflect a fundamental "animal logic" that, in our very large brains, guides all aspects of thought, including imagination. Further development of the argument requires Cooper to delve into the operation of the networks that implement the logic. He begins by dismissing the popular conceptualization of the brain as a computing system: "It is no more designed for logic or reason than the hand is designed to play the piano." The key to understanding how networks self-organize to produce thought is instead to be found in the mathematical rules that govern how individual synapses encode bits of information. Following on this deduction, we have an essay describing Cooper's astonishing success in deriving a set of these all-important rules from first principles and demonstrating that the rules actually operate in real brains. In all, I think Cooper does a brilliant job in making a case that a physical theory for thought is not only a realistic goal, but also one that may be within reach.

The third section of essays returns to the nature of science, but now with an interest in what, if any, are its limits. Are there problems that can't be solved? Cooper reminds us of essential questions (such as the chemical composition of stars) that once seemed out of reach only to fall eventually to new methods, the steady accumulation of data, or novel theory. Ever the optimist, he sees no reason why this won't continue with physicists eventually solving much-discussed problems that block the path to a grander theory. Notably, it is scientists who invented these problems; in the 19th century James Clerk Maxwell had no need to worry about quantum gravity, the nature of the Big Bang, or those nagging constants in the Standard Model of particle physics. Cooper argues that discovering new ways of seeing the world, whether prompted by experiment or derived from theory, is fundamental to science. And, he insists, the desire for new perspectives should be familiar to us because it arises from the same creative impulse that underlies great art or music. The Renaissance forever changed the discussion of what it means to be human—a revolutionary change in viewpoint not unlike the one that followed Einstein's reconceptualization of space and time.

But where is reality in this continuing replacement of one worldview for another? A key lies in the recognition that new theories do not invalidate older ones but instead go ever deeper into reality. Judging from its consequences, this process has been spectacularly successful: Newton will get a spaceship to Mars, but you'll need Erwin Schrödinger to download Netflix. And with these thoughts Cooper's essays return to the point that reality, at least as we know it, is a construction of the mind. Is it here, in attempting to generate a physical explanation for this most mysterious of phenomena, that science finally reaches its limits? Cooper doesn't think so. He sees mind and consciousness as one more—albeit particularly intimidating—example of a problem that only seems unsolvable. He even suggests steps toward an answer, including a surprising one: Don't hesitate to postulate unobservable mental activities. As described, this piece of advice flows naturally from Cooper's conceptualization of science.

And then at last there is the question of how bounded is the human mind, based, as it is, on slow hardware and erratic learning. To Cooper, these apparent weaknesses in fact result in perhaps the most extraordinary thing to be found in nature: "It pleases my natural optimism to be able to conclude that in this most important respect—imagination—there is no limit to human intellect. Our imagination is marvelously free, capable of any juxtaposition, unbounded by logic or experience." *Science and Human Experience* is a beautiful illustration of this conclusion.

15

Review: Michael S. Gazzaniga's *Tales from Both Sides of the Brain: A Life in Neuroscience*

By Theodor Landis, M.D.

Theodor Landis, M.D., is honorary professor of neurology at the University of Geneva, Switzerland, and former chair of the university's Department of Clinical Neurosciences. He is a behavioral neurologist with a special interest in cerebral hemispheric specialization and dual brain interaction. Landis studied medicine in Zürich and did his postgraduate and postdoctoral training in neurology and behavior neurology in Zürich, Bern, and Lausanne in Switzerland, as well as at the National Hospital, Queen Square, London, and the VA Medical Center's Aphasia Research Center in Boston. He is author or co-author of over 300 scientific peer-reviewed publications.

Editor's Note: Our brain has two hemispheres that specialize in different jobs—the right side processes spatial and temporal information, and the left side controls speech and language. How these two sides come together to create one mind is explained by pioneering neuroscientist Michael Gazzaniga in his new book, Tales from Both Sides of the Brain: A Life in Neuroscience *(Ecco/Harper Collins, 2015). Gazzaniga is director of the SAGE Center for the Study of the Mind at the University of California, Santa Barbara, and a Dana Alliance member.*

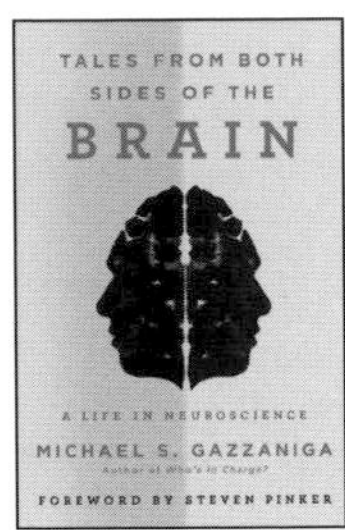

MICHAEL S. GAZZANIGA, one of the most important and leading figures in cognitive neuroscience, presents a brilliant and very lively autobiography that covers his contributions to one of the most puzzling and fascinating experiments in science: the disconnection of the two hemispheres of the brain in living patients.

As a young Ph.D. candidate in 1962, Gazzaniga was the right man at the right place: the California Institute of Technology, part of a group led by Roger Sperry, who later was awarded the Nobel Prize. Gazzaniga's description of his first experiment, which showed that the non-speaking, right cerebral hemisphere of his first "split-brain" patient understood the meaning of words, is as captivating as a thriller.

The "disconnectionist" approach to understanding brain functioning had begun with observations by the French neurologist Jules Dejerine, published in 1892, and by the German neurologist Hugo Liepmann, in 1900. These described the different functioning of the partly disconnected left and right cerebral hemispheres. Their work and others' paved the way for the Caltech experimenters' ultimate and complete "split" of the brain.

Gazzaniga begins his tale by recounting how, as an undergraduate at Dartmouth, he read of Sperry's work and became interested in the split-brain concept. He describes how patients with split brains differ from "normal" neurological patients, and how he used his knowledge of the split brain to explore a possible new window into the "unconscious."

Together with Bruce Volpe and Joseph LeDoux, a former student, Gazzaniga discovered in 1979 a way in which information that could not be accessed consciously could nevertheless influence seemingly conscious decisions. They thus opened a new field of cognitive research.

Gazzaniga describes the key experiments (why and how they came about, the fulfilled and unfulfilled expectations, and the differences between patients) and provides URLs to online videos of key experiments with some of his split-brain patients. These videos depict a condition that hasn't been fully explored, though it may be essential for understanding the self and the human mind.

Gazzaniga explains complex experiments and scientific discoveries in a simple, easy-to-understand manner. He makes important observations about the behavior of split-brain patients—observations that have never appeared before in the scientific literature. Many cognitive neuroscientists will surely use his observations to reflect on questions of neuroscience that have never been resolved.

Gazzaniga's account of the discovery of how split brains manage to "communicate" (by ingenious cross-cueing) is fascinating. It also demonstrates how easily a researcher, confronted with the peculiar scenario of a split brain, in search of unification, can miss the hemispheres' extra-callosal transfer of information and thus draw erroneous conclusions concerning mental integration.

Gazzaniga goes on to describe the amazing capacities of the non-speaking right cerebral hemisphere, and the wild confabulations of the speaking left hemisphere when asked to explain actions and decisions of its disconnected partner.

This book is also a tale of how science can be original and fun. For example, Gazzaniga in the course of his investigations converted a family van/trailer into a first-rank psychophysics-neuropsychology laboratory—allowing him and his team to test split-brain patients at their own homes.

Above all, Gazzaniga's story is about friendships, opportunities, academic and political encounters, and his many moves around the country to pursue intriguing opportunities. Gazzaniga's encounters with interesting and open-minded researchers, including the book's foreword contributor, Steven Pinker, read like a Who's Who of neuroscience.

Like some other famous cognitive neuroscientists of his era, Gazzaniga mentored a generation of new investigators in the field, many of whom have made huge contributions as well. The generous way in which he describes his relationships with former superiors, fellows, collaborators, postdocs, and students somehow defies his modest contention that he is an ungifted teacher.

The book is also a bit of a travelogue, as Gazzaniga traces his Italian heritage, great sense of hospitality, and love for food and wine. We learn that he often applied these traits in the name of science to organize lively scientific discussions in exotic locations, such as Venice, Tahiti, and Turkey. Out of these brainstorming conferences have come new ideas and experiments, a network of lifelong friends, and indeed an entire research community.

Gazzaniga's brilliant coining of the term "cognitive neuroscience" has helped create an intellectual home for researchers and clinicians trained in many other fields—from neuropsychology to philosophy and artificial intelligence.

Gazzaniga's book illuminates his flair for showmanship, which led him as a graduate student at Caltech to organize "on stage" political debates, on controversial topics, with politically opposed but witty commentators. Participants included TV personality Steve Allen, JFK biographer James MacGregor Burns, and leading conservative William F. Buckley, Jr. He notes that one debate even attracted Mr. and Mrs. Groucho Marx.

In 1989 Gazzaniga co-founded (with his wife of 33 years, Charlotte) the quarterly *Journal of Cognitive Neuroscience*, which has steadily risen in impact and is now a forum for work from all perspectives and subfields within cognitive neuroscience. His *Cognitive Neuroscience: A Reader* (Wiley, 2000) also has become the standard textbook in the field.

At the end of his autobiography Gazzaniga asks an important question: How does the brain enable mind?

He notes, in reference to the question, that there has been a revolution in neuroimaging and functional neuroimaging techniques, which now include CT scans; PET and SPECT scans with an ever increasing number of ligands; MRI scans with increasing magnetic field strengths and finer resolutions; MEG and three-dimensional EEG; deep brain recordings; and deep brain stimulations. He also cites an exponential increase in papers on cognitive neuroscience in animals, healthy humans, and patients.

However, despite this tremendous effort and progress, we still seem a long way from being able to explain how neurons produce "mind." Maybe the lifelong fascination of Gazzaniga with the patients who have disconnected brains is still the best way to test mind, but of course, only the divided mind.

16

Review:
Frances E. Jensen's *The Teenage Brain: A Neuroscientist's Survival Guide to Raising Adolescents and Young Adults*

Review by Marisa M. Silveri, Ph.D.

Marisa M. Silveri, Ph.D., is associate professor of psychiatry at the Harvard Medical School, adjunct assistant professor of psychiatry at the Boston University School of Medicine, and director of the Neurodevelopmental Laboratory on Addictions and Mental Health at McLean Hospital. Silveri uses magnetic resonance imaging (MRI) to study teen and emerging adult brain development, with a focus on identifying risk factors for substance abuse and psychiatric disorders. Silveri has received numerous awards and has more than 50 peer-reviewed scientific publications. She received a B.S. in biology/psychology from Union College and M.A. and Ph.D. degrees in behavioral neuroscience from the State University of New York at Binghamton.

Editor's Note: The unpredictable and sometimes incomprehensible moods and behaviors of a teenager can be a head-scratching mystery—especially to parents. Hormones, boredom, social media, peer pressure, and drugs and alcohol are just a few of the factors to consider. Frances E. Jensen, M.D., professor and chair of neurology at the University of Pennsylvania and the mother of two sons who are now in their 20s (along with Washington Post *health and science reporter and Pulitzer-Prize winner Amy Ellis Nutt), looks at the emerging science of the adolescent brain and provides advice based on Jensen's own research and experience as a single mother.*

MOST HUMANS, and to some extent other species, have one thing in common, regardless of age, race, gender identity, hair color, height, IQ, etc. We have all been teens—whether it was last week, in the 1980s, or in the 17th century. While the trials and tribulations facing teens may have changed dramatically over the millennia, we all have, or will have, traversed the chasm between childhood and maturity, collectively referred to as adolescence (from the Latin term adolescere, meaning "to grow up").

Author Frances E. Jensen, M.D., and co-writer Amy Ellis Nutt, are dedicated to raising public awareness regarding the teen brain. Jensen also—particularly relevant to the writing of a book on this subject—is the mother of two sons who graduated from adolescence. Publisher Harper touts *The Teenage Brain* as a neuroscientist's survival guide for understanding adolescence from multiple perspectives, and topics in this comprehensive tome include hormones, drugs and alcohol, historical perspectives, the impact of stress, mental illness, and matters legal and otherwise.

Jensen provides her sound scientific expertise and presents experimental brain data, as well as her firsthand practice of parenting through vignettes about her sons' sometimes questionable behavior—hair dye, a car crash, her response to a son having too much to drink as a college student. Jensen also presents humorous, cliché, and disheartening teen stories and testimonials from parents who have sought her advice. She also provides somewhat of an autobiographical account of her scientific career, which is interesting but perhaps not relevant to the main point of the book.

In terms of the science, Jensen writes in pseudo-lay terms to introduce the neurobiology of adolescent brain development. While having a background in neuroscience is not necessary to take full advantage of the advice offered, it is helpful. Early on, she emphasizes that until the past decade, the study of adolescence has been a relatively neglected research area. This is indeed true, and research in this area of brain science has been expanding precipitously, in part because of advances in technology that can reliably, and safely, provide a window into the human brain. Jensen and Ellis Nutt interpret the new data beautifully and accurately.

The book's overarching theme, presented from a variety of angles, is that adolescence is a period of unique opportunities and vulnerabilities. Each chapter includes historical perspectives on adolescence, from cultural to psychological to neurobiological. While historical information, including perspectives from different eras on the potency of drugs such as marijuana, the evolution of psychological studies on emotion and stress, and legal statutes for criminal culpability is interesting, the abundance of facts detracts somewhat from the neuroscience. Nonetheless, the authors successfully digest and interpret the brain science for parents and teens, with the goal of providing evidence and advice that will improve their navigation of this decade of unique life challenges. Jensen writes that their book "is about knowing [adolescents'] limits and what can be done to support them" and to provide "real data from real science journals"—and they indeed meet these goals. Figures reprinted with permission from previously published journal articles will be helpful to those who like having real data in hand, although these figures may lack the context necessary to have true utility to some lay readers.

The book is largely organized in bottom-up fashion, covering the more microscopic aspects of brain function and development, including emphasis on the importance of myelination and neuronal pruning in the teenage frontal lobe. Jensen points out that these processes contribute to a more efficient brain and largely underlie observed functional improvements during this age period. Further, she correctly underscores that teen brains are primed for knowledge, the main stage for the "opportunities" associated with adolescence. In contrast, Jensen writes that teenage "vulnerabilities" result from the well-documented evidence of increased teenage novelty seeking (i.e., risk taking), and with due diligence, she spends the remaining chapters of the book combing over studies on alcohol and drug effects.

The most useful information for parents may be the presentation of data on parental positions on alcohol and drug use, which have been shown to be a protective factor against future youth substance use and abuse. Stress, a ubiquitous experience that has transcended evolution, is also part of the equation. The book does a good job of describing the underlying biology of the damage stress can cause in teens, from the cellular level all the way to behavior, with notable impacts on memory and mental health.

Making this teenage guidebook especially timely is a focus on the obvious and the not-so-obvious perils of digital technology. Jensen keenly advocates for the formalization of Internet addictive disorder, which is not yet a diagnosis but certainly is a condition for which many adolescents, and adults, likely would meet diagnostic criteria.

The authors can't be blamed for not covering the impact of exercise or nutrition on teenage neurobiology, since those are areas where there are few existing data. And they do a good job of discussing where imaging has made an enormous impact in recent years: head injuries and mild traumatic brain injury, and the idea that the developing brain does not recover like that of an adult. Finally, although applicable to a somewhat smaller subset of the population, they discuss the juvenile justice system and offer the sound suggestion that local governments create rehabilitation and counseling programs for adolescents at risk, rather than build additional prison space and facilities.

The book closes by making parents aware that their offspring won't reach neurobiological adulthood until their mid-20s, rightly concluding that the period of opportunity and vulnerability extends beyond 18, an age once believed to be the gateway to adulthood.

Endnotes

1
Appraising the Risks of Reefer Madness

1. Andreasson S, Allebeck P, Engstrom A, Rydberg U. Cannabis and schizophrenia. A longitudinal study of Swedish conscripts. Lancet. 1987;2(8574):1483-6. Epub 1987/12/26.
2. Deglamorising cannabis. Lancet. 1995;346(8985):1241. Epub 1995/11/11.
3. Grech A, Van Os J, Jones PB, Lewis SW, Murray RM. Cannabis use and outcome of recent onset psychosis. European psychiatry : the journal of the Association of European Psychiatrists. 2005;20(4):349-53. Epub 2005/07/16.
4. Moore TH, Zammit S, Lingford-Hughes A, Barnes TR, Jones PB, Burke M, et al. Cannabis use and risk of psychotic or affective mental health outcomes: a systematic review. Lancet. 2007;370(9584):319-28. Epub 2007/07/31.
5. Zammit S, Moore TH, Lingford-Hughes A, Barnes TR, Jones PB, Burke M, et al. Effects of cannabis use on outcomes of psychotic disorders: systematic review. The British journal of psychiatry : the journal of mental science. 2008;193(5):357-63. Epub 2008/11/04.
6. Foti DJ, Kotov R, Guey LT, Bromet EJ. Cannabis use and the course of schizophrenia: 10-year follow-up after first hospitalization. The American journal of psychiatry. 2010;167(8):987-93. Epub 2010/05/19.
7. Arseneault L, Cannon M, Poulton R, Murray R, Caspi A, Moffitt TE. Cannabis use in adolescence and risk for adult psychosis: longitudinal prospective study. BMJ. 2002;325(7374):1212-3. Epub 2002/11/26.
8. Fergusson DM, Horwood LJ, Ridder EM. Tests of causal linkages between cannabis use and psychotic symptoms. Addiction. 2005;100(3):354-66. Epub 2005/03/01.
9. McGrath J, Welham J, Scott J, Varghese D, Degenhardt L, Hayatbakhsh MR, et al. Association between cannabis use and psychosis-related outcomes using sibling pair analysis in a cohort of young adults. Archives of general psychiatry. 2010;67(5):440-7. Epub 2010/03/03.
10. van Os J, Bak M, Hanssen M, Bijl RV, de Graaf R, Verdoux H. Cannabis use and psychosis: a longitudinal population-based study. American journal of epidemiology. 2002;156(4):319-27. Epub 2002/08/16.
11. Henquet C, Krabbendam L, Spauwen J, Kaplan C, Lieb R, Wittchen HU, et al. Prospective cohort study of cannabis use, predisposition for psychosis, and psychotic symptoms in young people. BMJ. 2005;330(7481):11. Epub 2004/12/03.
12. Stefanis NC, Delespaul P, Henquet C, Bakoula C, Stefanis CN, Van Os J. Early adolescent cannabis exposure and positive and negative dimensions of psychosis.

Addiction. 2004;99(10):1333-41. Epub 2004/09/17.
13. Ferdinand RF, Sondeijker F, van der Ende J, Selten JP, Huizink A, Verhulst FC. Cannabis use predicts future psychotic symptoms, and vice versa. Addiction. 2005;100(5):612-8. Epub 2005/04/26.
14. Weiser M, Knobler HY, Noy S, Kaplan Z. Clinical characteristics of adolescents later hospitalized for schizophrenia. American journal of medical genetics. 2002;114(8):949-55. Epub 2002/11/29.
15. Rossler W, Hengartner MP, Angst J, Ajdacic-Gross V. Linking substance use with symptoms of subclinical psychosis in a community cohort over 30 years. Addiction. 2012;107(6):1174-84. Epub 2011/12/14.
16. Griffith-Lendering MF, Wigman JT, Prince van Leeuwen A, Huijbregts SC, Huizink AC, Ormel J, et al. Cannabis use and vulnerability for psychosis in early adolescence--a TRAILS study. Addiction. 2013;108(4):733-40. Epub 2012/12/12.
17. Zammit S, Allebeck P, Andreasson S, Lundberg I, Lewis G. Self reported cannabis use as a risk factor for schizophrenia in Swedish conscripts of 1969: historical cohort study. BMJ. 2002;325(7374):1199. Epub 2002/11/26.
18. Manrique-Garcia E, Zammit S, Dalman C, Hemmingsson T, Andreasson S, Allebeck P. Cannabis, schizophrenia and other non-affective psychoses: 35 years of follow-up of a population-based cohort. Psychological medicine. 2012;42(6):1321-8. Epub 2011/10/18.
19. Niemi-Pynttari JA, Sund R, Putkonen H, Vorma H, Wahlbeck K, Pirkola SP. Substance-induced psychoses converting into schizophrenia: a register-based study of 18,478 Finnish inpatient cases. The Journal of clinical psychiatry. 2013;74(1):e94-9. Epub 2013/02/20.
20. Callaghan RC, Cunningham JK, Allebeck P, Arenovich T, Sajeev G, Remington G, et al. Methamphetamine use and schizophrenia: a population-based cohort study in California. The American journal of psychiatry. 2012;169(4):389-96. Epub 2011/12/24.
21. Casadio P, Fernandes C, Murray RM, Di Forti M. Cannabis use in young people: the risk for schizophrenia. Neuroscience and biobehavioral reviews. 2011;35(8):1779-87. Epub 2011/05/03.
22. Macleod J, Oakes R, Copello A, Crome I, Egger M, Hickman M, et al. Psychological and social sequelae of cannabis and other illicit drug use by young people: a systematic review of longitudinal, general population studies. Lancet. 2004;363(9421):1579-88. Epub 2004/05/18.
23. Arseneault L, Cannon M, Witton J, Murray RM. Causal association between cannabis and psychosis: examination of the evidence. The British journal of psychiatry : the journal of mental science. 2004;184:110-7. Epub 2004/02/03.
24. Radhakrishnan R, Wilkinson ST, D'Souza DC. Gone to Pot - A Review of the Association between Cannabis and Psychosis. Frontiers in psychiatry. 2014;5:54. Epub 2014/06/07.
25. Di Forti M, Iyegbe C, Stilo SA, Murray RM, et al. The proportion of first episode psychosis attributable to cannabis use. Lancet Psychiatry. In Press
26. Kedzior KK, Laeber LT. A positive association between anxiety disorders and cannabis use or cannabis use disorders in the general population--a meta-analysis of 31 studies. BMC psychiatry. 2014;14:136. Epub 2014/06/03.
27. Manrique-Garcia E, Zammit S, Dalman C, Hemmingsson T, Allebeck P. Can-

nabis use and depression: a longitudinal study of a national cohort of Swedish conscripts. BMC psychiatry. 2012;12:112. Epub 2012/08/18.

28. Henquet C, Di Forti M, Morrison P, Kuepper R, Murray RM. Gene-environment interplay between cannabis and psychosis. Schizophrenia bulletin. 2008;34(6):1111-21. Epub 2008/08/30.
29. Caspi A, Moffitt TE, Cannon M, McClay J, Murray R, Harrington H, et al. Moderation of the effect of adolescent-onset cannabis use on adult psychosis by a functional polymorphism in the catechol-O-methyltransferase gene: longitudinal evidence of a gene X environment interaction. Biological psychiatry. 2005;57(10):1117-27. Epub 2005/05/04.
30. Di Forti M, Iyegbe C, Sallis H, Kolliakou A, Falcone MA, Paparelli A, et al. Confirmation that the AKT1 (rs2494732) genotype influences the risk of psychosis in cannabis users. Biological psychiatry. 2012;72(10):811-6. Epub 2012/07/27.
31. Colizzi M, Di Forti M, Gaughran F, Stilo S et al Interaction between a functional variant in DRD2 and cannabis use in psychosis. Schizophrenia Bulletin.In revision
32. Potter DJ, Clark P, Brown MB. Potency of delta 9-THC and other cannabinoids in cannabis in England in 2005: implications for psychoactivity and pharmacology. Journal of forensic sciences. 2008;53(1):90-4. Epub 2008/02/19.
33. Pijlman T, Rigter SM, Hoek I,Goldsmidt MJ, and Niesink RJ Strong increase in total delta-THC in cannabis preparations sold in Dutch coffee shops. Addiction Biology (June 2005) 10, 171 – 180.
34. Moreau, J. J. Hashish and Mental Illness (Raven, New York, 1973).
35. Morrison PD, Zois V, McKeown DA, Lee TD, Holt DW, Powell JF, et al. The acute effects of synthetic intravenous Delta9-tetrahydrocannabinol on psychosis, mood and cognitive functioning. Psychological medicine. 2009;39(10):1607-16. Epub 2009/04/02.
36. Englund A, Morrison PD, Nottage J, Hague D, Kane F, Bonaccorso S, et al. Cannabidiol inhibits THC-elicited paranoid symptoms and hippocampal-dependent memory impairment. J Psychopharmacol. 2013;27(1):19-27. Epub 2012/10/09.
37. Leweke FM, Piomelli D, Pahlisch F, Muhl D, Gerth CW, Hoyer C, et al. Cannabidiol enhances anandamide signaling and alleviates psychotic symptoms of schizophrenia. Translational psychiatry. 2012;2:e94. Epub 2012/07/27.
38. Di Forti M, Morgan C, Dazzan P, Pariante C, Mondelli V, Marques TR, et al. High-potency cannabis and the risk of psychosis. The British journal of psychiatry : the journal of mental science. 2009;195(6):488-91. Epub 2009/12/02.
39. Di Forti M, Sallis H, Allegri F, Trotta A, Ferraro L, Stilo SA, et al. Daily use, especially of high-potency cannabis, drives the earlier onset of psychosis in cannabis users. Schizophrenia bulletin. 2014;40(6):1509-17. Epub 2013/12/19.
40. Morgan CJ, Curran HV. Effects of cannabidiol on schizophrenia-like symptoms in people who use cannabis. The British journal of psychiatry : the journal of mental science. 2008;192(4):306-7. Epub 2008/04/02.
41. European Monitoring Centre for Drugs and Drug Addiction. New developments in Europe's cannabis market. 2014
42. Winstock AR, Barratt MJ. Synthetic cannabis: a comparison of patterns of use and effect profile with natural cannabis in a large global sample. Drug and alcohol dependence. 2013;131(1-2):106-11. Epub 2013/01/08.
43. Papanti D, Schifano F, Botteon G, Bertossi F, Mannix J, Vidoni D, et al. "Spi-

ceophrenia": a systematic overview of "spice"-related psychopathological issues and a case report. Human psychopharmacology. 2013;28(4):379-89. Epub 2013/07/25.

44. Spaderna M, Addy PH, D'Souza DC. Spicing things up: synthetic cannabinoids. Psychopharmacology. 2013;228(4):525-40. Epub 2013/07/10.
45. Khan SS, Secades-Villa R, Okuda M, Wang S, Perez-Fuentes G, Kerridge BT, et al. Gender differences in cannabis use disorders: results from the National Epidemiologic Survey of Alcohol and Related Conditions. Drug and alcohol dependence. 2013;130(1-3):101-8. Epub 2012/11/28.
46. European Monitoring Centre for Drugs and Drug Addiction (EMCDDA). Drug Treatment Overview for Netherlands. Lisbon: EMCDDA; 2013. Available at: http://www.emcdda.europa.eu/data/treatment-overviews/Netherlands (accessed 23 July 2014). (Archived athttp://www.webcitation.org/6S4yjPY59 on 25 August 2014).
47. Hall W. What has research over the past two decades revealed about the adverse health effects of recreational cannabis use? Addiction. 2015;110(1):19-35. Epub 2014/10/08.
48. Solowij N. Cannabis and Cognitive Functioning Published by Cambridge University Press, 2006 ISBN 10: 0521591147 ISBN 13: 9780521591140.
49. Meier MH, Caspi A, Ambler A, Harrington H, Houts R, Keefe RS, et al. Persistent cannabis users show neuropsychological decline from childhood to midlife. Proceedings of the National Academy of Sciences of the United States of America. 2012;109(40):E2657-64. Epub 2012/08/29.
50. Pope HG, Jr., Gruber AJ, Hudson JI, Cohane G, Huestis MA, Yurgelun-Todd D. Early-onset cannabis use and cognitive deficits: what is the nature of the association? Drug and alcohol dependence. 2003;69(3):303-10. Epub 2003/03/14.
51. Silins E, Horwood LG, Patton GC, Fergusson DM et al Young adult sequelae of adolescent cannabis use:an integrative analysis. Lancet Psychiatry 2014; 1: 286–93.
52. Gilman JM, Kuster JK, Lee S, Lee MJ, Kim BW, Makris N, et al. Cannabis use is quantitatively associated with nucleus accumbens and amygdala abnormalities in young adult recreational users. The Journal of neuroscience : the official journal of the Society for Neuroscience. 2014;34(16):5529-38. Epub 2014/04/18.
53. Yucel M, Solowij N, Respondek C, Whittle S, Fornito A, Pantelis C, et al. Regional brain abnormalities associated with long-term heavy cannabis use. Archives of general psychiatry. 2008;65(6):694-701. Epub 2008/06/04.
54. Filbey FM, Aslan S, Calhoun VD, Spence JS, Damaraju E, Caprihan A, et al. Long-term effects of marijuana use on the brain. Proceedings of the National Academy of Sciences of the United States of America. 2014;111(47):16913-8. Epub 2014/11/12.
55. Johnston LD, O'Malley PM, Miech RA, et al. Monitoring the Future: national survey results on drug use, 1975-2013 —overview, key findings on adolescent drug use. Ann Arbor: Institute for Social Research, University of Michigan, 2014 (http://monitoringthefuture.org/pubs/monographs/mtf-overview2013.pdf).
56. Drug misuse: Findings from the 2013/14 Crime Survey for England and Wales. Home Office. UK Government.
57. UK Drug Policy Commission. A Fresh Approach to Drugs, 2012
58. Sevigny EL, Pacula RL, Heaton P. The effects of medical marijuana laws on

potency. The International journal on drug policy. 2014;25(2):308-19. Epub 2014/02/08.

59. United Nations. Commission on Narcotic Drugs. Report to the 56th session. World situation on Drug Abuse.Vienna 11-15th March. 2013.

2
Why Inspiring Stories Make Us React: The Neuroscience of Narrative

1. P.J. Zak, R. Kurzban, and W.T. Matzner, "The Neurobiology of Trust," Annals of the New York Academy of Sciences 1032(2004): 224-227.
2. P.J. Zak, R. Kurzban, and W.T. Matzner, "Oxytocin is Associated with Human Trustworthiness," Hormones and Behavior 48(2005): 522-527.
3. V. B. Morhenn, J. W. Park, E. Piper, and P.J. Zak, "Monetary Sacrifice Among Strangers is Mediated by Endogenous Oxytocin Release after Physical Contact. Evolution and Human Behavior 29(2008): 375–383.
4. C.S. Carter, and E.B. Keverne, "The Neurobiology of Social Affiliation and Pair Bonding," in Hormones, Brain and Behavior, ed. Donald Pfaff, (San Diego: Academic Press, 2002): 299–337.
5. C. T. Wotjak, J. Ganster, G. Kohl, F. Holsboer, R. Landgraf, and M.Engelmann, "Dissociated Central and Peripheral Release of Vasopressin, but not Oxytocin, in Response to Repeated Swim Stress: New Insights into the Secretory Capacities of Peptidergic Neurons," Neuroscience 85(1998): 1209–1222 doi: 10.1016/S0306-4522(97)00683-0.
6. I. D. Neumann, "Brain Oxytocin: A Key Regulator of Emotional and Social Behaviours in Both Females and Males," Journal of Neuroendocrinology 20(2008): 858–865 doi: 10.1111/j.1365-2826.2008.01726.x.
7. P.J. Zak, "The Moral Molecule: The Source of Love and Prosperity," (New York: Dutton, 2012).
8. Zak, P. J., Stanton, A. A. & Ahmadi, S. 2007. Oxytocin increases generosity in humans. Public Library of Science ONE, 2(11).
9. M. Kosfeld, M. Heinrichs, P. J. Zak, U.Fischbacher, and E. Fehr, "Oxytocin Increases Trust in Humans," Nature 435:2(2005): 673-676.
10. P.J. Zak, and J.A. Barraza, "The Neurobiology of Collective Action," Frontiers in Neuroscience. 7(2013):211 doi: 10.3389/fnins.2013.00211.
11. J. A. Barraza, and P. J. Zak, "Empathy Toward Strangers Triggers Oxytocin Release and Subsequent Generosity." Annals of the New York Academy of Sciences 1167(2009): 182-189.
12. P-Y. Lin, N.S. Grewal, C. Morin, W.D. Johnson, and P.J. Zak, "Oxytocin Increases the Influence of Public Service Advertisements," PLoS ONE 8:2(2013) doi: 10.1371/journal.pone.0056934.
13. J.A. Barraza, V. Alexander, L.E. Beavin, E.T. Terris, E.T., and P.J. Zak, "The Heart of the Story: Peripheral Physiology During Narrative Exposure Predicts Charitable Giving," Biological Psychology (2015): in press.
14. R.J. Gerrig, "Experiencing Narrative Worlds: On the Psychological Activities of Reading," (New Haven: Yale University Press, 1993).
15. D.Kahneman, and R.H. Thaler, "Anomalies: Utility Maximization and Experi-

enced Utility," Journal of Economic Perspectives 20:1(2006):221–234.
16. B.K. Bracken, V. Alexander, P.J. Zak, V. Romero, and J.A. Barraza, "Physiological Synchronization is Associated with Narrative Emotionality and Subsequent Behavioral Response," in Foundations of Augmented Cognition. Advancing Human Performance and Decision-Making through Adaptive Systems: 8th International Conference,, eds., D.D. Schmorrow and C.M. Fidopiastis (Berlin: Springer, 2014) 3–13.
17. S. Gimbel, J. Kaplan, M. Immordino-Yang, Tipper, C. A. Gordon, M. Dehghani, S. Sagae, H. Damasio, A. Damasio, "Neural Response to Narratives Framed with Sacred Values," Abstract, Annual meeting of the Society for Neuroscience, (San Diego, 2013).

4

The Darkness Within: Individual Differences in Stress

1. H. Selye, "Selye's Guide to Stress Research," Van Nostrand Reinhold, New York, 1990.
2. S.R. Burchfield, "The stress response: a new perspective," Psychosomatic Medicine 41 (1979): 661-672.
3. W. Vale, J. Spiess, C. Rivier, and J. Rivier, "Characterization of a 41-residue ovine hypothalamic peptide that stimulates the secretion of corticotropin and b-endorphin," Science 213 (1981): 1394-1397.
4. R.P. Rao, S. Anilkumar, B.S. McEwen, and S. Chattarji. "Glucocorticoids protect against the delayed behavioral and cellular effects of acute stress on the amygdala." Biological Psychiatry 72 (2012) 466-475.
5. G.F. Koob, and M. Le Moal, "Addiction and the brain antireward system," Annual Review of Psychology 59 (2008): 29-53.
6. J.E. LeDoux, "Emotion circuits in the brain," Annual Review of Neuroscience 23 (2000): 155-184.
7. G.F. Alheid, and L. Heimer, "New perspectives in basal forebrain organization of special relevance for neuropsychiatric disorders: the striatopallidal, amygdaloid, and corticopetal components of substantia innominata," Neuroscience 27 (1988): 1-39.
8. S.M. Reynolds, S. Geisler, A. Bérod, and D.S. Zahm. "Neurotensin antagonist acutely and robustly attenuates locomotion that accompanies stimulation of a neurotensin-containing pathway from rostrobasal forebrain to the ventral tegmental area." European Journal of Neuroscience 24 (2006): 188-196.
9. L.M. Shin, and I. Liberzon, "The neurocircuitry of fear, stress, and anxiety disorders," Neuropsychopharmacology 35 (2010): 169-191.
10. G.F. Koob, "Neuroadaptive mechanisms of addiction: studies on the extended amygdala," European Neuropsychopharmacology 13 (2003): 442-452.
11. R.E. Sutton, G.F. Koob, M. Le Moal, J. Rivier, and W. Vale, "Corticotropin releasing factor produces behavioural activation in rats," Nature 297 (1982): 331-333
12. G.F. Koob, and E.P. Zorrilla, "Neurobiological mechanisms of addiction: focus on corticotropin-releasing factor," Current Opinion in Investigational Drugs 11 (2010): 63-71.
13. A. Van't Veer A, and W.A. Carlezon, Jr., "Role of kappa-opioid receptors in stress

and anxiety-related behavior," Psychopharmacology 229 (2013): 435-452.

14. S.J. Watson, H. Khachaturian, H. Akil, D.H. Coy, and A. Goldstein, "Comparison of the distribution of dynorphin systems and enkephalin systems in brain," Science 218 (1982): 1134-1136.
15. J. H. Fallon, and F.M. Leslie, "Distribution of dynorphin and enkephalin peptides in the rat brain," Journal of Comparative Neurology 249 (1986): 293-336.
16. G.F. Koob, "The dark side of emotion: the addiction perspective," European Journal of Pharmacology (2014) in press.
17. P. Ouimette, J. Read, and P.J. Brown, "Consistency of retrospective reports of DSM-IV Criterion A traumatic stressors among substance use disorder patients," Journal of Traumatic Stress 18 (2005): 43–51.
18. C. Herry, F. Ferraguti, N. Singewald, J.J. Letzkus, I. Ehrlich, and A. Luthi, "Neuronal circuits of fear extinction," European Journal of Neuroscience 31 (2010): 599-612.
19. R. Yehuda, "Status of glucocorticoid alterations in post-traumatic stress disorder," Annals of the New York Academy of Sciences 1179 (2009): 56-69.
20. M. Van Zuiden, E. Geuze, H.L. Willermen, E. Vermetten, M. Maas, C.J. Heijnen, and A. Kavelaars, "Pre-existing high glucocorticoid receptor number predicting development of posttraumatic stress symptoms after military deployment," American Journal of Psychiatry 168 (2011): 89e96.
21. V.B Risbrough, and M.B. Stein, "Role of corticotropin releasing factor in anxiety disorders: a translational research perspective," Hormones and Behavior 50 (2006): 550-561.
22. D.G. Baker, S.A. West, W.E. Nicholson, N.N. Ekhator, J.W. Kasckow, K.K. Hill, A.B. Bruce, D.N. Orth, and T.D. Geracioti, "Serial CSF CRH levels and adrenocortical activity in combat veterans with posttraumatic stress disorder," American Journal of Psychiatry, 156 (1999): 585-588.
23. R.H. Pietrzak, M. Naganawa, Y. Huang, S. Corsi-Travali, M.Q. Zheng, M.B. Stein, S. Henry, K. Lim, J. Ropchan, S.F. Lin, R.E. Carson, and A. Neumeister, "Association of in vivo k-opioid receptor availability and the transdiagnostic dimensional expression of trauma-related psychopathology," JAMA Psychiatry 71 (2014): 1262-1270.
24. J.P. Hayes, S.M. Hayes, and A.M. Mikedis, "Quantitative meta-analysis of neural activity in posttraumatic stress disorder," Biology of Mood and Anxiety Disorders 2 (2012): 9.
25. S.L. Rauch, L.M. Shin, and E.A. Phelps, "Neurocircuitry models of posttraumatic stress disorder and extinction: human neuroimaging research: past, present, and future," Biological Psychiatry 60 (2006): 376-382.
26. R.K. Pitman, A.M. Rasmusson, K.C. Koenen, L.M. Shin, S.P. Orr, M.W. Gilbertson, M.R. Milad, and I. Liberzon, "Biological studies of post-traumatic stress disorder," Nature Reviews Neuroscience 13 (2012): 769-787.
27. M.R. Milad, G.J. Quirk, R.K. Pitman, S.P. Orr, B. Fischl, and S.L. Rauch, "A role for the human dorsal anterior cingulate cortex in fear expression," Biological Psychiatry 62 (2007a): 1191-1194.
28. M.R. Milad, C.I. Wright, S.P. Orr, R.K. Pitman, G.J. Quirk, and S.L. Rauch, "Recall of fear extinction in humans activates the ventromedial prefrontal cortex and hippocampus in concert," Biological Psychiatry 62 (2007b): 446-454.
29. G.F. Koob, and M. Le Moal, "Drug abuse: hedonic homeostatic dysregulation,"

Science 278 (1997): 52-58.
30. G.F. Koob, and N.D. Volkow, "Neurocircuitry of addiction," Neuropsychopharmacology Reviews 35 (2010): 217-238.
31. F.J. Vaccarino, H.O. Pettit, F.E. Bloom, and G.F. Koob. "Effects of intracerebroventricular administration of methyl naloxonium chloride on heroin self-administration in the rat." Pharmacology Biochemistry and Behavior 23 (1985): 495-498.
32. N.D. Volkow, G.J. Wang, F. Telang, J.S. Fowler, J. Logan , M. Jayne, Y. Ma, K. Pradhan, and C. Wong, "Profound decreases in dopamine release in striatum in detoxified alcoholics: possible orbitofrontal involvement," Journal of Neuroscience 27 (2007): 12700-12706.
33. J.M. Mitchell, J.P. O'Neil, M. Janabi, S.M. Marks, W.J. Jagust, and H.L. Fields, "Alcohol consumption induces endogenous opioid release in the human orbitofrontal cortex and nucleus accumbens," Science Translational Medicine 4 (2012): 116ra6.
34. S.B. Sparber, and D.R. Meyer, "Clonidine antagonizes naloxone-induced suppression of conditioned behavior and body weight loss in morphine-dependent rats," Pharmacology Biochemistry and Behavior 9 (1978): 319-325.
35. G.F. Koob, T.L. Wall, and F.E. Bloom, "Nucleus accumbens as a substrate for the aversive stimulus effects of opiate withdrawal," Psychopharmacology 98 (1989): 530-534.
36. G.F. Koob, and M. Le Moal, "Drug addiction, dysregulation of reward, and allostasis," Neuropsychopharmacology, 24 (2001): 97-129.
37. W.A. Carlezon, Jr., E.J. Nestler, and R.L. Neve, "Herpes simplex virus-mediated gene transfer as a tool for neuropsychiatric research," Critical Reviews in Neurobiology 14 (2000): 47-67.
38. H.A. Baldwin, S. Rassnick, J. Rivier, G.F. Koob, and K.T. Britton, "CRF antagonist reverses the "anxiogenic" response to ethanol withdrawal in the rat," Psychopharmacology 103 (1991): 227-232.
39. E. Merlo-Pich, M. Lorang, M. Yeganeh, F. Rodriguez de Fonseca, J. Raber, G.F. Koob, and F. Weiss, "Increase of extracellular corticotropin-releasing factor-like immunoreactivity levels in the amygdala of awake rats during restraint stress and ethanol withdrawal as measured by microdialysis," Journal of Neuroscience 15 (1995): 5439-5447.
40. C.K. Funk, L.E. O'Dell, E.F. Crawford, and G.F. Koob, "Corticotropin-releasing factor within the central nucleus of the amygdala mediates enhanced ethanol self-administration in withdrawn, ethanol-dependent rats," Journal of Neuroscience 26 (2006): 11324-11332.
41. L.F. Vendruscolo, E. Barbier, J.E. Schlosburg, K.K. Misra, T. Whitfield, Jr., M.L. Logrip, C.L. Rivier, V. Repunte-Canonigo, E.P. Zorrilla, P.P. Sanna, M. Heilig, and G.F. Koob. "Corticosteroid-dependent plasticity mediates compulsive alcohol drinking in rats." Journal of Neuroscience 32 (2012) 7563-7571.
42. B.M. Walker, E.P. Zorrilla, and G.F. Koob, "Systemic k-opioid receptor antagonism by nor-binaltorphimine reduces dependence-induced excessive alcohol self-administration in rats," Addiction Biology 16 (2010): 116-119.
43. S. Wee, L. Orio, S. Ghirmai, J.R. Cashman, and G.F. Koob, "Inhibition of kappa opioid receptors attenuated increased cocaine intake in rats with extended access to cocaine," Psychopharmacology 205 (2009): 565-575.

44. T.W Whitfield, Jr., J. Schlosburg, S. Wee, L. Vendruscolo, A. Gould, O. George, Y. Grant, S. Edwards, E. Crawford, and G. Koob, "Kappa opioid receptors in the nucleus accumbens shell mediate escalation of methamphetamine intake," Journal of Neuroscience (2014) in press.
45. J.E. Schlosburg, T.W. Whitfield, Jr., P.E. Park, E.F. Crawford, O. George, L.F. Vendruscolo, and G.F. Koob. "Long-term antagonism of opioid receptors prevents escalation of and increased motivation for heroin intake." Journal of Neuroscience 33 (2013): 19384-19392.
46. G.F. Koob, "Negative reinforcement in drug addiction: the darkness within," Current Opinion in Neurobiology 23 (2013): 559-563.
47. O. George, C. Sanders, J. Freiling, E. Grigoryan, C.D. Vu, S. Allen, E. Crawford, C.D. Mandyam, and G.F. Koob, "Recruitment of medial prefrontal cortex neurons during alcohol withdrawal predicts cognitive impairment and excessive alcohol drinking," Proceedings of the National Academy of Sciences of the United States of America 109 (2012): 18156-18161.
48. J.L. Perry, J.E. Joseph, Y. Jiang, R.S. Zimmerman, T.H. Kelly, M. Darna, P. Huettl, L.P. Dwoskin, and M.T. Bardo. "Prefrontal cortex and drug abuse vulnerability: translation to prevention and treatment interventions." Brain Research Reviews 65 (2011): 124-149.
49. A. Etkin, K.C. Klemenhagen, J.T. Dudman, M.T. Rogan, R. Hen, E.R. Kandel, and J. Hirsch, "Individual differences in trait anxiety predict the response of the basolateral amygdala to unconsciously processed fearful faces," Neuron 44 (2004): 1043-1055.
50. A.R. Childress, P.D. Mozley, W. McElgin, J. Fitzgerald, M. Reivich, and C.P. O'Brien, "Limbic activation during cue-induced cocaine craving," American Journal of Psychiatry 156 (1999): 11-18.
51. A. Sekiguchi, M. Sugiura, Y. Taki, Y. Kotozaki, R. Nouchi, H. Takeuchi, T. Araki, S. Hanawa, S. Nakagawa, C.M. Miyauchi, A. Sakuma, and R. Kawashima, "Brain structural changes as vulnerability factors and acquired signs of post-earthquake stress," Molecular Psychiatry 18 (2013): 618-623.
52. K. Felmingham, A. Kemp, L. Williams, P. Das, G. Hughes, A. Peduto, and R. Bryant, "Changes in anterior cingulate and amygdala after cognitive behavior therapy of posttraumatic stress disorder," Psychological Science 18 (2007): 127-129.
53. L.M. Shin, N.B, Lasko, M.L. Macklin, R.D. Karpf, M.R. Milad, S.P. Orr, J.M. Goetz, A.J. Fischman, S.L. Rauch, and R.K. Pitman, "Resting metabolic activity in the cingulate cortex and vulnerability to posttraumatic stress disorder," Archives of General Psychiatry 66 (2009): 1099-1107.
54. M. Uddin, A.E. Aiello, D.E. Wildman, K.C. Koenen, G. Pawelec, R. de Los Santos, E. Goldmann, and S. Galea, "Epigenetic and immune function profiles associated with posttraumatic stress disorder," Proceedings of the National Academy of Sciences of the United States of America 107 (2010): 9470-9475.
55. S.C. Pandey, R. Ugale, H. Zhang, L. Tang, and A. Prakash, "Brain chromatin remodeling: a novel mechanism of alcoholism," Journal of Neuroscience 28 (2008): 3729-3737.
56. K.C. Koenen, M. Uddin, S.C. Chang, A.E. Aiello, D.E. Wildman, E. Goldmann, and S. Galea, "SLC6A4 methylation modifies the effect of the number of traumatic events on risk for posttraumatic stress disorder," Depression and Anxiety 28 (2011): 639-647.

57. S. Moonat, A.J. Sakharkar, H. Zhang, L. Tang, and S.C. Pandey, "Aberrant histone deacetylase2-mediated histone modifications and synaptic plasticity in the amygdala predisposes to anxiety and alcoholism," Biological Psychiatry 73 (2013): 763-773.

5
Tau-er of Power

1. Gsponer J, Futschik ME, Teichmann SA, Babu MM. Tight regulation of unstructured proteins: from transcript synthesis to protein degradation. Science 2008;322(5906):1365-8.
2. Babu MM, van der Lee R, de Groot NS, Gsponer J. Intrinsically disordered proteins: regulation and disease. Curr Opin Struct Biol 2011;21(3):432-40.
3. Santa-Maria I, Alaniz ME, Renwick N, Cela C, Fulga TA, Van Vactor D, Tuschl T, Clark LN, Shelanski ML, McCabe BD and others. Dysregulation of microRNA-219 promotes neurodegeneration through post-transcriptional regulation of tau. J Clin Invest 2015;125(2):681-6.
4. Baas PW, Pienkowski TP, Kosik KS. Processes induced by tau expression in Sf9 cells have an axon-like microtubule organization. J Cell Biol 1991;115(5):1333-44.
5. Halfmann R, Wright JR, Alberti S, Lindquist S, Rexach M. Prion formation by a yeast GLFG nucleoporin. Prion 2012;6(4):391-9.
6. Lee VM, Giasson BI, Trojanowski JQ. More than just two peas in a pod: common amyloidogenic properties of tau and alpha-synuclein in neurodegenerative diseases. Trends Neurosci 2004;27(3):129-34.
7. Clavaguera F, Akatsu H, Fraser G, Crowther RA, Frank S, Hench J, Probst A, Winkler DT, Reichwald J, Staufenbiel M and others. Brain homogenates from human tauopathies induce tau inclusions in mouse brain. Proc Natl Acad Sci U S A 2013;110(23):9535-40.
8. Sanders DW, Kaufman SK, DeVos SL, Sharma AM, Mirbaha H, Li A, Barker SJ, Foley AC, Thorpe JR, Serpell LC and others. Distinct tau prion strains propagate in cells and mice and define different tauopathies. Neuron 2014;82(6):1271-88.
9. Liu L, Drouet V, Wu JW, Witter MP, Small SA, Clelland C, Duff K. Trans-synaptic spread of tau pathology in vivo. PLoS One 2012;7(2):e31302.
10. de Calignon A, Polydoro M, Suarez-Calvet M, William C, Adamowicz DH, Kopeikina KJ, Pitstick R, Sahara N, Ashe KH, Carlson GA and others. Propagation of tau pathology in a model of early Alzheimer's disease. Neuron 2012;73(4):685-97.
11. Kfoury N, Holmes BB, Jiang H, Holtzman DM, Diamond MI. Trans-cellular propagation of Tau aggregation by fibrillar species. J Biol Chem 2012;287(23):19440-51.
12. Nelson R, Sawaya MR, Balbirnie M, Madsen AO, Riekel C, Grothe R, Eisenberg D. Structure of the cross-beta spine of amyloid-like fibrils. Nature 2005;435(7043):773-8.
13. Sievers SA, Karanicolas J, Chang HW, Zhao A, Jiang L, Zirafi O, Stevens JT, Munch J, Baker D, Eisenberg D. Structure-based design of non-natural amino-acid inhibitors of amyloid fibril formation. Nature 2011;475(7354):96-100.
14. Carrettiero DC, Hernandez I, Neveu P, Papagiannakopoulos T, Kosik KS. The

cochaperone BAG2 sweeps paired helical filament- insoluble tau from the microtubule. J Neurosci 2009;29(7):2151-61.

15. Gamblin TC, Berry RW, Binder LI. Modeling tau polymerization in vitro: a review and synthesis. Biochemistry 2003;42(51):15009-17.
16. Levine ZA, Larini L, LaPointe NE, Feinstein SC, Shea JE. Regulation and aggregation of intrinsically disordered peptides. Proc Natl Acad Sci U S A 2015;112(9):2758-63.
17. Walker LC, Diamond MI, Duff KE, Hyman BT. Mechanisms of protein seeding in neurodegenerative diseases. JAMA Neurol 2013;70(3):304-10.
18. Seeley WW, Crawford RK, Zhou J, Miller BL, Greicius MD. Neurodegenerative diseases target large-scale human brain networks. Neuron 2009;62(1):42-52.
19. Raj A, Kuceyeski A, Weiner M. A network diffusion model of disease progression in dementia. Neuron 2012;73(6):1204-15.
20. Menkes-Caspi N, Yamin HG, Kellner V, Spires-Jones TL, Cohen D, Stern EA. Pathological tau disrupts ongoing network activity. Neuron 2015;85(5):959-66.

6
New Movement in Neuroscience: A Purpose-Driven Life

1. V Frankl. Man's Search for Meaning: Simon and Schuster; 1946.
2. J Crumbaugh, L Maholick. "An experimental study in existentialism: The psychometric approach to Frankl's concept of noogenic neurosis". Journal of Clinical Psychology. (1964): 20(2): 200-207.
3. Alzheimer's Association. What is Alzheimer's Disease. alz.org. 2015. http://www.alz.org/alzheimers_disease_what_is_alzheimers.asp .
4. P Boyle, A Buchman, L Barnes and D Bennett. "Effect of a Purpose in Life on Risk of Incident Alzheimer Disease and Mild Cognitive Impairment in Community-Dwelling Older Persons". Arch Gen Psychiatry. (2010): 304-310.
5. P Boyle, A Buchman, R Wilson, L Yu, J Schneider and D Bennett. "Effect of Purpose in Life on the Relation Between Alzheimer Disease Pathologic Changes on Cognitive Function in Advanced Age". Arch Gen Psychiatry. (2012): 499-506.
6. W Harvey. Exercitatio Anatomica de Motu Cordis et Sanguinis in Animalibus (An Anatomical Exercise Concerning the Motion of the Heart and Blood in Animals); 1628.
7. American Stroke Association. About Stroke. American Stroke Association. 2015. http://www.strokeassociation.org/STROKEORG/AboutStroke/About-Stroke_UCM_308529_SubHomePage.jsp .
8. E Kim, J Sun, N Park and C Peterson. "Purpose in Life and Reduced Incidence of Stroke in Older Adults: The Health and Retirement Study". Journal of Psychosomatic Research. (2013): 74: 427-432.
9. E Kim, J Sun, N Park, L Kubzansky, C Peterson. "Purpose in Life and Reduced Risk of Myocardial Infarction Among Older US Adults with Coronary Heart Disease: A Two Year Followup". J Behav Med. (2013): 36: 124-133
10. M Koizumi, H Ito, Y Kaneko and Y Motohashi. "Effect of Having a Sense of Purpose in Life on the Risk of Death from Cardiovascular Diseases". J Epidemiology. (2008): 18(5): 191-196.
11. SM Lucas, N Rothwell and R Gibson. "The role of inflammation in CNS injury and disease". British Journal of Pharmacology. (2006): S232-S240.

12. N Rohleder. "Stimulation of Systemic Low-Grade Inflammation by Psychosocial Stress". Psychosomatic Medicine. (2014): 181-189.
13. E Friedman, M Hayney, G Love, B Singer and C Ryff. "Plasma interleukin-6 and soluble IL-6 receptors are associated with psychological well-being in aging women". Health Psychology. (2007): 305-313.
14. B Fredrickson, K Grewen, K Coffrey, S Algoe and A Firestine. "A functional genomic perspective on human well-being". PNAS. (2013): 13684-13689.
15. R Baumeister, K Vohs, J Aaker and E Garbinsky. "Some Key Differences between a Happy Life and a Meaningful Life". The Journal of Positive Psychology. (2013): 8(6): 505-516.
16. P Hedburg, Y Gustafson, L Alex and C Brulin. "Depression in relation to purpose in life among a very old population: A five-year follow-up study". Aging and Mental Health. (2010): 14(6): 757-763.
17. E Telzer, A Fuligni, M Lieberman and A Galvan. "Neural sensitivity to eudaimonic and hedonic rewards differentially predict adolescent depressive symptoms over time". PNAS. (2014): 111(8): 6600-6605.
18. M Blazek, T Besta. "Self-Concept Clarity and Religious Orientations: Prediction of Purpose in Life and Self-Esteem". Journal of Religion and Health. (2012): 51(3): 947-960.
19. A Levit, S Licina. How the Recession Shaped Millennial and Hiring Manager Attitudes about Millennials' Future Careers 2011.
20. S Thompson, J Pitts. "Factors relating to a person's ability to find meaning after a diagnosis of cancer". Journal of Psychosocial Oncology. (1994): 11(3): 1-21.
21. C Holahan, C Holahan and R Suzuki. "Purposiveness, physical activity and perceived health in cardiac patients". Disability and Rehabilitation. (2008): 30(23): 1772-1778. V Frankl. Man's Search for Meaning: Simon and Schuster; 1946.

7
Schizophrenia: Hope on the Horizon

1. Murray CJ, Vos T, Lozano R, Naghavi M, Flaxman AD, Michaud C, et al. Disability-adjusted life years (DALYs) for 291 diseases and injuries in 21 regions, 1990–2010: a systematic analysis for the Global Burden of Disease Study 2010. Lancet. 2012;380(9859):2197-223.
2. Sullivan PF. The Psychiatric GWAS Consortium: big science comes to psychiatry. Neuron. 2010;68:182-6.
3. Schizophrenia Working Group of the Psychiatric Genomics Consortium. Biological insights from 108 schizophrenia-associated genetic loci. Nature. 2014;511:421-7.
4. Levinson DF, Duan J, Oh S, Wang K, Sanders AR, Shi J, et al. Copy number variants in schizophrenia: Confirmation of five previous findings and new evidence for 3q29 microdeletions and VIPR2 duplications. Am J Psychiatry. 2011;168:302-16.
5. Purcell SM, Moran JL, Fromer M, Ruderfer D, Solovieff N, Roussos P, et al. A polygenic burden of rare disruptive mutations in schizophrenia. Nature. 2014;506:185-90.
6. Owen MJ. New approaches to psychiatric diagnostic classification. Neuron. 2014;84(3):564-71.

7. Sullivan PF, Daly MJ, O'Donovan M. Genetic architectures of psychiatric disorders: the emerging picture and its implications. Nature Reviews Genetics. 2012;13:537-51.

8
The Holy Grail of Psychiatry

1. Styron W. Darkness Visible: A Memoir of Madness. 1990. Random House. New York.
2. US Burden of Disease Collaborators. The State of US Health, 1990-2010 Burden of Diseases, Injuries, and Risk Factors. JAMA (2013): 591-606.
3. Ozomaro U, Wahlestedt C, Nemeroff CB. Personalized Medicine in Psychiatry: Problems and Promises. BMC Medicine (2013): 11:132.
4. Ozomaro U, Nemeroff CB, Wahlestedt C. Personalized Medicine and Psychiatry: Dream or Reality? Psychiatric Times (2013): 26-29.
5. Nemeroff CB. Psychopharmacology and the Future of Personalized Treatment. Depression and Anxiety (2014): 906-908.
6. Alhajji L, Nemeroff CB. Personalized Medicine and Mood Disorders. Psychiatric Clinics of North America (2015). In press.
7. McGrath CL, Kelley ME, Holtzheimer PE, Dunlop BW, Craighead WE, Franco AR, Craddock C, Mayberg H. Toward a Neuroimaging Treatment Selection Biomarker for Major Depressive Disorder. JAMA (2013): 821-829.
8. Nanni V, Uher R, Danese A. Childhood Maltreatment Predicts Unfavorable Course of Illness and Treatment Outcome in Depression: A Meta-Analysis. American Journal of Psychiatry (2012): 141-151.
9. Kaiser RH, Andrews-Hanna JR, Wager TD, Pizzagalli DA. Large-Scale Network Dysfunction in Major Depressive Disorder. JAMA Psychiatry (2015): 603-611.

9
No End in Sight: The Abuse of Prescription Narcotics

1. Way, E.L. (1982). History of opiate use in the Orient and the United States. Annals of the New York Academy of Sciences, 398: 12-23.
2. Wright, A. D. (1968). The history of opium. Medical & biological illustration, 18(1), 62-70.
3. Griffin, N., Khoshnood, K. (2010). Opium trade, insurgency, and HIV/AIDS in Afghanistan: relationships and regional consequences. Asia-Pacific Journal of Public Health, 22(3 suppl), 159S-167S.
4. Phillips, W. J., & Currier, B. L. (2004). Analgesic pharmacology: II. Specific analgesics. Journal of the American Academy of Orthopaedic Surgeons, 12(4), 221-233.
5. Fields, H. L., & Margolis, E. B. (2015). Understanding opioid reward. Trends in Neurosciences, 38(4), 217-225.
6. Brecher, E. M. "The Harrison Narcotic Act (1914)," Chapter 8 in The Consumers Union Report on Licit and Illicit Drugs. Retrieved from http://www.druglibrary.org/schaffer/library/studies/cu/cu8.html.
7. Wolf, P.L. (2010). Hector Berlioz and other famous artists with opium abuse. Frontiers of Neurology and Neuroscience, 27: 84-91.
8. Osler, W. (n.d.). A brief history of opium: the plant of joy. BLTC Research. Re-

trieved from http://www.opiates.net.
9. NIDA's addiction research center (ARC) 60th anniversary: History of the addiction research center (Nov. 1995). National Institute on Drug Abuse. Retrieved from http://archives.drugabuse.gov/NIDA_Notes/NNVol10N6/ARCHistory.html.
10. Institute of Medicine (US) Committee on Advancing Pain Research, Care, and Education (2011). Relieving Pain in America: A Blueprint for Transforming Prevention, Care, Education, and Research. National Academies Press. As referenced in a 2011 IOM report.
11. Aronoff, G. M. (2000). Opioids in chronic pain management: is there a significant risk of addiction? Current Review of Pain, 4(2), 112-121.
12. Gregg, J. (2015). A Startling Injustice: Pain, Opioids, and Addiction. Annals of Internal Medicine, 162(9), 651-652.
13. Benzinger, D. P., Miotto, J., Grandy, R. P., Thomas, G. B., Swanson, R. E., & Fitzmartin, R. D. (1997). A pharmacokinetic/pharmacodynamic study of controlled-release oxycodone. Journal of Pain and Symptom Management, 13(2), 75-82.
14. Purdue Pharma, execs to pay $634.5 million fine in OxyContin case (2007). Associated Press. Retrieved from http://www.cnbc.com/id/18591525.
15. Cicero, T. J., & Ellis, M. S. (2015). Abuse-Deterrent Formulations and the Prescription Opioid Abuse Epidemic in the United States: Lessons Learned From OxyContin. JAMA Psychiatry, 72(5), 424-430.
16. Cicero, T. J., Ellis, M. S., & Surratt, H. L. (2012). Effect of abuse-deterrent formulation of OxyContin. New England Journal of Medicine, 367(2), 187-189.
17. Edlund, M. J., Forman-Hoffman, V. L., Winder, C. R., Heller, D. C., Kroutil, L. A., Lipari, R. N., & Colpe, L. J. (2015). Opioid abuse and depression in adolescents: results from the National Survey on Drug Use and Health. Drug and Alcohol Dependence.
18. Manchikanti, L., Fellows, B., & Ailinani, H. (2010). Therapeutic use, abuse, and nonmedical use of opioids: a ten-year perspective. Pain Physician, 13, 401-435.
19. Dart, R. C., Surratt, H. L., Cicero, T. J., Parrino, M. W., Severtson, S. G., Bucher-Bartelson, B., & Green, J. L. (2015). Trends in opioid analgesic abuse and mortality in the United States. New England Journal of Medicine, 372(3), 241-248.
20. Cicero, T. J., Ellis, M. S., Surratt, H. L., & Kurtz, S. P. (2014). The changing face of heroin use in the United States: a retrospective analysis of the past 50 years. JAMA Psychiatry, 71(7), 821-826.
21. Maxwell, J. C. (2015). The Pain Reliever and Heroin Epidemic in the United States: Shifting Winds in the Perfect Storm. Journal of Addictive Diseases, (just-accepted), 00-00.
22. Cicero, T.J., Kuehn, B. M. (2014). Driven by prescription drug abuse, heroin use increases among suburban and rural whites. JAMA, 312(2), 118-119.
23. Jones, C. M. (2013). Heroin use and heroin use risk behaviors among nonmedical users of prescription opioid pain relievers–United States, 2002–2004 and 2008–2010. Drug and Alcohol Dependence, 132(1), 95-100.

10
The Binge and the Brain

1. Luce KH, Crowther JH, Pole M. Eating Disorder Examination Questionnaire (EDE-Q): norms for undergraduate women. Int J Eat Disord. 2008;41(3):273-276.
2. Association AP. Diagnostic and Statistical Manual of Mental Disorders (DSM-5®). Arlington: American Psychiatric Publishing; 2013.
3. Stunkard A. A description of eating disorders in 1932. Am J Psychiatry. 1990;147(3):263-268.
4. Lillis J, Hayes SC, Levin ME. Binge eating and weight control: the role of experiential avoidance. Behav Modif. 2011;35(3):252-264.
5. McClure SM, Laibson DI, Loewenstein G, Cohen JD. Separate neural systems value immediate and delayed monetary rewards. Science. 2004;306(5695):503-507.
6. Mobbs O, Iglesias K, Golay A, Van der Linden M. Cognitive deficits in obese persons with and without binge eating disorder. Investigation using a mental flexibility task. Appetite. 2011;57(1):263-271.
7. Balodis IM, Molina ND, Kober H, Worhunsky PD, White MA, Rajita S, Grilo CM, Potenza MN. Divergent neural substrates of inhibitory control in binge eating disorder relative to other manifestations of obesity. Obesity (Silver Spring). 2013;21(2):367-377.
8. Schienle A, Schafer A, Hermann A, Vaitl D. Binge-eating disorder: reward sensitivity and brain activation to images of food. Biol Psychiatry. 2009;65(8):654-661.
9. Balodis IM, Kober H, Worhunsky PD, White MA, Stevens MC, Pearlson GD, Sinha R, Grilo CM, Potenza MN. Monetary reward processing in obese individuals with and without binge eating disorder. Biol Psychiatry. 2013;73(9):877-886.
10. Wang GJ, Geliebter A, Volkow ND, Telang FW, Logan J, Jayne MC, Galanti K, Selig PA, Han H, Zhu W, Wong CT, Fowler JS. Enhanced striatal dopamine release during food stimulation in binge eating disorder. Obesity (Silver Spring). 2011;19(8):1601-1608.
11. Atiye M, Miettunen J, Raevuori-Helkamaa A. A meta-analysis of temperament in eating disorders. Eur Eat Disord Rev. 2015;23(2):89-99.
12. Pearson CM, Zapolski TC, Smith GT. A longitudinal test of impulsivity and depression pathways to early binge eating onset. Int J Eat Disord. 2014.
13. Fairburn CG, Cooper Z, Doll HA, O'Connor ME, Bohn K, Hawker DM, Wales JA, Palmer RL. Transdiagnostic cognitive-behavioral therapy for patients with eating disorders: a two-site trial with 60-week follow-up. Am J Psychiatry. 2009;166(3):311-319.
14. Eldredge KL, Stewart Agras W, Arnow B, Telch CF, Bell S, Castonguay L, Marnell M. The effects of extending cognitive-behavioral therapy for binge eating disorder among initial treatment nonresponders. Int J Eat Disord. 1997;21(4):347-352.
15. Safer DL, Robinson AH, Jo B. Outcome from a randomized controlled trial of group therapy for binge eating disorder: comparing dialectical behavior therapy adapted for binge eating to an active comparison group therapy. Behav Ther. 2010;41(1):106-120.
16. Wallace LM, Masson PC, Safer DL, von Ranson KM. Change in emotion regulation during the course of treatment predicts binge abstinence in guided self-help

dialectical behavior therapy for binge eating disorder. J Eat Disord. 2014;2(1):35.
17. Vocks S, Tuschen-Caffier B, Pietrowsky R, Rustenbach SJ, Kersting A, Herpertz S. Meta-analysis of the effectiveness of psychological and pharmacological treatments for binge eating disorder. Int J Eat Disord. 2010;43(3):205-217.
18. McElroy SL, Hudson JI, Mitchell JE, Wilfley D, Ferreira-Cornwell MC, Gao J, Wang J, Whitaker T, Jonas J, Gasior M. Efficacy and safety of lisdexamfetamine for treatment of adults with moderate to severe binge-eating disorder: a randomized clinical trial. JAMA Psychiatry. 2015;72(3):235-246.
19. Boutelle KN, Liang J, Knatz S, Matheson B, Risbrough V, Strong D, Rhee KE, Craske MG, Zucker N, Bouton ME. Design and implementation of a study evaluating extinction processes to food cues in obese children: the Intervention for Regulations of Cues Trial (iROC). Contemp Clin Trials. 2015;40:95-104.

11
Failure to Replicate: Sound the Alarm

1. M. Bakker, A. van Dijk, and J.M. Wicherts, "The Rules of Game Called Psychological Science," Perspectives on Psychological Science 7 (2012): 543-54.
2. D. Fanelli, ""Positive" Results Increase Down the Hierarchy of the Sciences," PLoS ONE 5 (2010): e10068.
3. S.E. Maxwell, "The Persistence of Underpowered Studies in Psychological Research: Causes, Consequences, and Remedies." Psychological Methods, 9 (2004): 147-163.
4. J.P. Ioannidis, M. Munafò, P. Fusar-Poli, B.A. Nosek, S. David, "Publication and Other Reporting Biases in Cognitive Sciences: Detection, Prevalence and Prevention" Trends in Cognitive Sciences 18 (2014): 235-41.
5. K.S. Button, J.P. Ioannidis, C. Mokrysz, B.A. Nosek, J. Flint, E.S. Robinson, and M.R. Munafò, "Power Failure: Why Small Sample Size Undermines the Reliability of Neuroscience," Nature Reviews Neuroscience 14 (2013): 365-76.
6. M. Makel, J. Plucker, and B. Hegarty, "Replications in Psychology Research: How Often Do they really Occur?" Perspectives on Psychological Science 6 (2012): 537-42.
7. J.P. Ioannidis, Why Most Published Research Findings are False. PLoS Medicine 2 (2005): e124.
8. K. Fiedler, "Voodoo Correlations are Everywhere. Not Only in Neuroscience," Perspectives on Psychological Science 6 (2011): 163-171.
9. B.A. Nosek, and Y. Bar-Anan, "Scientific Utopia: I. Opening Scientific Communication," Psychological Inquiry 23 (2012): 217-43.
10. J.P. Ioannidis, "Scientific Inbreeding and Same-Team Replication," Journal of Psychosomatic Research 73 (2012): 408-10.
11. Open Science Collaboration, "Estimating the Reproducibility of Psychological Science," Science 349 (2015): aac4716.
12. L.E. Williams, J.A. Bargh, J. A, "Keeping One's Distance: The Influence of Spatial Distance Cues on Affect and Evaluation," Psychological Science 19 (2008): 302-8.
13. P. Fischer, T. Greitemeyer, D. Frey, "Self-regulation and Selective Exposure: The Impact of Depleted Self-regulation Resources on Confirmatory Information Processing," Journal of Personality and Social Psychology 94(2008): 382-395.

14. H. Evanschitzky, C. Baumgarth, R. Hubbard, and J.S. Armstrong, "Replication Research's Disturbing Trend," Journal of Business Research 60 (2007): 411-5.
15. R. Hubbard, and J.S. Armstrong, "Replication and Extensions in Marketing: Rarely Published but Quite Contrary," International Journal of Research in Marketing 11 (1994): 233-48.
16. C.W. Kelly, L.J. Vhase, and R.K. Tucker, "Replication in Experimental Communication Research: an Analysis," Human Communication Research 5 (1999): 338-42.
17. C.G. Begley, and L.M. Ellis, "Drug Development: Raise Standards for Preclinical Cancer Research," Nature 483 (2012): 531-3.
18. F. Prinz, T. Schlange, and K. Asadullah, "Believe it or not: How Much can we Rely on Published Data on Potential Drug Targets?" Nature Reviews Drug Discovery 10 (2011): 712-713.
19. J.P. Ioannidis, R. Tarone, and J.K. McLaughlin, "The False Positive to False Negative Ratio in Epidemiologic Studies," Epidemiology 22 (2011): 450-6.
20. J.P. Ioannidis, "Why Science is not Necessarily Self-correcting," Perspectives in Psychological Sciences 7 (2012): 645-54.
21. J.P. Ioannidis, "How to Make More Published Research True," PLoS Medicine 11 (2005): e1001747.
22. D.L. Donoho, A. Maleki, I.U. Rahman, M. Shahram, and V. Stodden, V, "Reproducibility Research in Computational Harmonic Analysis," Computing, Science, & Engineering 11 (2009): 8-18.
23. J.M. Wicherts, D. Borsboom, J. Kats, and D. Molenaar, "The Poor Availability of Psychological Research Data for Reanalysis," American Psychologist 61 (2006): 726–728.
24. J.W. Neuliep, and R. Crandall, "Editorial Bias Against Replication Research," Journal of Social Behavior and Personality 5 (1990): 85–90.
25. J.W. Neuliep, and R. Crandall, "Reviewer Bias Against Replication Research," Journal of Social Behavior and Personality 8 (1993): 21–29.

Financial Disclosure: Meta-Research Innovation Center at Stanford University (METRICS) is funded by a grant from the Laura and John Arnold Foundation. The work of John Ioannidis is also supported by an unrestricted gift from Sue and Bob O'Donnell.

12
Cognitive Skills and the Aging Brain: What to Expect

1. Hartshorne, JK and Germine, LT. When does cognitive functioning peak? The asynchronous rise and fall of different cognitive abilities across the life span. Psychological Science, 2015; 26:433-443.
2. Insurance Institute for Highway Safety (IIHS). Fatality facts 2013, Older people. Arlington (VA): IIHS; 2014. [cited 2015 Mar 26]. Available from URL: http://www.iihs.org/iihs/topics/t/older-drivers/fatalityfacts/older-people/2013
3. PsychCorp. Wechsler Memory Scale-Fourth Edition (WMS-IV) Technical and Interpretative Manual. San Antonio, TX, Pearson, 2009.
4. Denburg, NL, Cole, CA, Hernandez, M, et al. The orbitofrontal cortex, real-world decision-making, and normal aging. Annals of the New York Academy of Sciences, 2007; 1121: 480-498.

5. Hess, TM and Queen, TL. Aging influences on judgment and decision processes: Interactions between ability and experience. In Verhaeghen, P, and Hertzog, C (Eds) The Oxford Handbook of Emotion, Social Cognition, and Problem Solving in Adulthood. Oxford University Press, 2014, pp. 238-255.
6. Tucker AM and Stern, Y. Cognitive reserve in aging. Current Alzheimer Research, 2011; 8: 354-360.
7. Underwood, E. Starting young. Science, 2014; 346 no. 6209: 568-571.
8. Dodge, HH, Ybarra, O, Kaye, JA. Tools for advancing research into social networks and cognitive function in older adults. International Psychogeriatrics, 2014, 26: 533-539.
9. Buchman, AS, Yu, L, Wilson, RS, Boyle, PA, Schneider, JA, Bennett, DA. Brain pathology contributes to simultaneous change in physical frailty and cognition in old age. The Journals of Gerontology. Series A, Biological Sciences and Medical Sciences. 2014; 69: 1536-1544.

Index

C

D

R

S

T

V

W

Z